The Celestial Mechanics

I0407478

Navigating the Cosmos

NAGINDERPAL SINGH

The Celestial Mechanics

By Naginderpal Singh

ISBN: 9798865308935

Cover Design: Naginderpal Singh

Interior Design: Naginderpal Singh

Publisher: The Enlightenment

Published in United States of America

First Edition: 2023

DEDICATED TO

I am dedicating this book to my parents Sardar Narinder Singh and Late Sardarni Gurjeet Kaur

FOREWORD

In "The Celestial Mechanics: Navigating the Cosmos," the reader is about to embark on an astonishing trip through the vast complexity of our cosmic surroundings. This book is a monument to the everlasting human curiosity that pulls us ever farther into the infinite expanse of space.

Celestial mechanics, the study of understanding and forecasting the motions of celestial bodies, is a field that has uncovered the mysteries of our solar system and beyond. Through the years, visionaries like Kepler, Newton, and Euler have led the way, and now, this book gives a thorough and accessible reference to their enormous achievements.

From the exquisite simplicity of Kepler's equations to the mind-bending accuracy of current astrodynamics, the chapters inside are tailored to educate both the interested beginner and the seasoned astrophysicist. With a rigorous combination of theory and practice, this book establishes a basis for appreciating the complicated dance of planets, moons, and stars.

In these pages, readers will journey through subjects ranging from the principles of orbital dynamics to the mystical worlds of chaos theory and resonance. They will study the mathematical miracles that control our spacecraft's excursions across the universe, from accurate interplanetary transfers to exquisite rendezvous in the emptiness.

The exploration doesn't stop there. Venturing into the future, we explore into the boundaries of celestial mechanics, where interstellar transport and quantum effects await our intellectual grasp. These are the challenges that call the next generation of spacefarers, the visionaries and pioneers who will drive humanity to new worlds and unveil the secrets of the cosmos.

As we flip each page, it is my aim that readers will not only obtain a thorough comprehension of celestial physics but also nurture a better appreciation for the marvels that surround us. For in our attempt to traverse the universe, we build a link with something much bigger than ourselves, a cosmic tapestry that weaves together the tales of innumerable celestial bodies and the civilizations that look upon them.

So, to the intrepid reader, prepare for an adventure that transcends the boundaries of space and time. Allow the wisdom inside these pages to be your compass as you begin on your own trip across the universe. For in this trip, we discover not just solutions to scientific concerns but also the inexhaustible inspiration that motivates our collaborative study of the cosmos.

Narinder Singh

PREFACE

In the magnificent fabric of the cosmos, the dance of heavenly bodies has grabbed the imagination of mankind for millennia. The beautiful arcs of planets, the fascinating trails of comets, and the distant twinkling of stars have stirred our interest, prompting us to understand the mysteries that govern their motions. It is this dance, this exquisite choreography of the universe, that forms the crux of our research in this book: "The Celestial Mechanics: Navigating the universe."

As we stand on the verge of an age defined by extraordinary technological growth and a revitalized thirst for space exploration, the study of celestial physics has never been more important or interesting. From the beautiful rules of Newton to the complexity of interplanetary trajectories, this book goes on a trip through the mathematical and physical foundations that enable us to map journeys across the cosmos.

Through these pages, we shall cover the historical landmarks that have formed our knowledge of celestial physics. From Kepler's groundbreaking rules of planetary motion to the present applications of perturbation theory, each chapter offers a stepping stone toward a greater grasp of the universe. We will investigate the subtleties of orbital dynamics, dig into the problems of spacecraft navigation, and journey into the world of interplanetary transfers.

Yet, this book is more than a simple compilation of mathematical formulae and scientific truths. It is an invitation to wonder at the beauty and intricacy of the cosmos, to admire the genius of those who have gone before us, and to contemplate the possibilities that lie ahead. It is a monument to the human spirit of exploration, to our constant quest of knowledge, and to our potential to transcend the bounds of our home planet.

While this voyage may be challenging, with ideas both deep and profound, it is our goal that it will be a rewarding one. Whether you are an aspiring astrophysicist, a seasoned aerospace engineer, or just an avid stargazer, we encourage you to embark on this voyage across the universe. Together, let us travel the routes of celestial physics, building a closer relationship with the beauties that adorn our night sky.

In concluding, we convey our profound appreciation to the pioneers of celestial mechanics whose discoveries opened the path for our expedition. We also applaud the persistent work of individuals who continue to push the limits of space science and technology, lighting new horizons in our understanding of the cosmos.

May this book serve as a guiding light on your cosmic voyage, creating a love for the universe that will continue to blaze brilliantly for centuries to come.

Naginderpal Singh

ACKNOWLEDGMENTS

I would want to convey my heartfelt appreciation to my sister, Birinder Pal Kaur, whose unfailing support and encouragement played a key part in the production of this book. Her trust in me and her inexhaustible tolerance throughout the writing process were crucial. Birinder's insightful thoughts and thoughtful suggestions improved the material and gave a new perspective to the complicated topic of celestial mechanics.

Her continual interest for research and adventure has been a fountain of inspiration. Her enthusiasm for the universe and her inquisitive personality have been a motivating factor behind my own interest for this topic. I am extremely blessed to have such a loyal and supporting sister.

I am appreciative to Birinder for her numerous hours of conversations, her rigorous editing, and her ongoing conviction in the significance of spreading information about the marvels of our universe. This book serves as a testimony to her impact and contribution.

With deepest appreciation,

Naginderpal Singh

CONTENTS

Chapter 1: The Dance of the Celestial Bodies

1.1 Introduction to Celestial Mechanics

"Introduction to Celestial Mechanics" serves as a basic literature in the topic of astrodynamics, providing readers with a complete grasp of the laws guiding the motion of celestial bodies within our solar system. Authored by J. M. A. Danby, the book presents a well-balanced combination of historical background, theoretical foundations, and practical applications, making it accessible to both beginners and seasoned specialists in the area.

One significant quality of the book is in its historical perspective, which provides readers with a contextual foundation for comprehending the development of celestial mechanics. By diving into the contributions of luminaries like Johannes Kepler, Isaac Newton, and Pierre-Simon Laplace, Danby elucidates the intellectual path that resulted in the contemporary knowledge of celestial dynamics. This historical backdrop not only offers richness to the subject matter but also emphasizes the tremendous human effort that led to our present understanding of celestial physics.

The presentation of essential topics such as Newton's laws of motion and the universal law of gravity is another significant part of Danby's work. The author provides these essential ideas with a clarity that enables understanding, especially for individuals without a strong experience in physics. By stressing the mathematical rigor needed for celestial mechanics, Danby offers readers with the skills essential for further examination of the subject area.

Moreover, "Introduction to Celestial Mechanics" shines in its description of practical applications. Through the explication of Kepler's laws of planetary motion and the notion of orbital components, Danby sets the framework for comprehending the motion of celestial bodies in space. These notions are not only theoretical constructions; they have wide-ranging applications in space exploration, satellite navigation, and mission planning, making this book a useful resource for experts in the aerospace industry.

The book's strength resides not only in its substance but also in its instructional approach. Danby follows a logical succession of themes, ensuring that readers are directed from basic ideas to more sophisticated notions. Additionally, the inclusion of illustrated examples and activities improves understanding and encourages readers to apply theoretical information to actual settings.

In conclusion, "Introduction to Celestial Mechanics" by J. M. A. Danby stands as an invaluable resource in the area of astrodynamics. Its historical background, clear presentation of basic ideas, and practical applications make it a significant source for students, researchers, and practitioners in the subject of celestial mechanics. By bridging the gap between theory and practice, Danby's work not only transmits information but also instills a greater appreciation for the complex dance of celestial bodies that molds our view of the universe.

1.2 Historical Overview

The historical backdrop of celestial mechanics is a trip through millennia of human curiosity and scientific study. This basic subject, which digs into the motion of celestial bodies, has strong origins in ancient civilizations. Early astronomers, such as the

Babylonians and Greeks, examined the motions of stars and planets, striving to comprehend the patterns that ruled the night sky. Their findings set the basis for what would eventually become the discipline of celestial mechanics.

One of the crucial points in the history of celestial mechanics occurred with the work of Johannes Kepler in the 17th century. Kepler's Laws of Planetary Motion altered our knowledge of how planets circle the Sun. Through thorough studies, he established three essential laws that characterized the elliptical motions of planets around the Sun, replacing the previously accepted belief in perfect circles. This innovation established a mathematical foundation for forecasting planetary placements, marking a key turning point in celestial mechanics.

The 17th century also saw the important contributions of Sir Isaac Newton. His formulation of the Universal Law of Gravitation and his three Laws of Motion supplied the theoretical basis for celestial mechanics. Newton's revelation that the same force driving the fall of an apple also regulated the motion of planets ushered the discipline into a new era. It allowed for exact predictions of planetary orbits and even permitted the discovery of Neptune, a monument to the efficacy of mathematical models in celestial mechanics.

The 18th and 19th centuries witnessed the refining and spread of celestial mechanics. Pioneering astronomers like Pierre-Simon Laplace and Joseph-Louis Lagrange devised mathematical ways to manage complicated gravitational interactions, opening the door for the study of multi-body systems. Laplace's work on perturbation theory, which analyzes the effects of minor gravitational forces on orbits, proved important in properly forecasting the motions of celestial bodies.

The 20th century ushered in a new age of celestial mechanics with the advent of space exploration. Engineers and scientists, propelled by the goal to reach the moon and beyond, relied extensively on the concepts of celestial mechanics to calculate trajectories and perform exact operations. The Apollo missions, for instance, displayed the astounding practical application of celestial mechanics, displaying humanity's capacity to navigate across the expanse of space.

In more recent times, celestial mechanics has found new applications in the study of exoplanetary systems and in preparing expeditions to distant celestial worlds. The accuracy expected by current space missions needs a sophisticated grasp of orbital mechanics and gravitational interactions. As we look to the future, celestial mechanics will definitely continue to play a critical part in our attempts to explore and navigate the universe, underlying the next generation of space exploration.

In summary, the historical review of celestial mechanics illustrates a remarkable growth of human intelligence and scientific success. From the early observations of ancient astronomers to the mathematical rigor of Kepler and Newton, and the practical applications in space travel, this field has influenced our knowledge of the universe and enhanced our capacity to navigate and explore the expanse of space. Today, celestial mechanics stands as a tribute to the continuing curiosity and creativity of mankind in the face of the cosmic unknown.

1.3 The Celestial Sphere and Coordinate Systems

"The Celestial Sphere and Coordinate Systems" is a key idea in celestial mechanics and astronomy. It gives a simple model for viewing the locations and motions of celestial objects from an observer's viewpoint on Earth. The notion of the celestial sphere posits that all stars and celestial bodies are positioned on an imagined, endlessly vast sphere around the Earth. This sphere is a crucial tool for astronomers since it enables them to project the three-dimensional locations of stars onto a two-dimensional plane, simplifying the process of identifying and monitoring celestial objects.

One fundamental feature of the celestial sphere is the formation of coordinate systems. These systems give a framework for defining the locations of celestial objects relative to an observer on Earth. The two basic coordinate systems used in astronomy are the equatorial and ecliptic coordinate systems. The equatorial system is based on the projection of the Earth's equator onto the celestial sphere, providing lines of right ascension (analogous to longitude on Earth) and declination (equivalent to latitude). This approach offers a stable reference frame against which the locations of stars and other celestial objects may be measured.

On the other hand, the ecliptic coordinate system is aligned with the plane of Earth's orbit around the Sun, known as the ecliptic plane. It comprises of celestial longitude (measured along the ecliptic) and celestial latitude (measured perpendicular to the ecliptic). This method is especially essential for monitoring the locations of planets and other objects inside our solar system.

Understanding these coordinate systems is vital for celestial navigation, allowing astronomers to properly find and anticipate the motions of celestial bodies. It supplies the foundation for a broad variety of astronomical studies, from following the movements of planets to determining the locations of distant galaxies.

Moreover, the celestial sphere and coordinate systems play a crucial role in the development of astrodynamics and space missions. Engineers and scientists utilize these principles to calculate and forecast the orbits of artificial satellites and interplanetary missions. By using the concepts of celestial mechanics, they can calculate trajectories, make maneuvers, and assure exact location of spacecraft in the vastness of space.

In conclusion, "The Celestial Sphere and Coordinate Systems" is a core idea in celestial mechanics, providing astronomers and astrodynamics professionals with a coherent framework for detecting and monitoring celestial objects. The creation of coordinate systems, notably the equatorial and ecliptic systems, gives an organized technique to measuring the locations of stars, planets, and other celestial bodies. This information provides the foundation for a broad variety of astronomical observations and is vital in planning and performing space missions. It constitutes a cornerstone in the investigation and comprehension of the universe.

Chapter 2: Newton's Laws and Orbital Dynamics

2.1 Newton's Laws of Motion

Newton's Laws of Motion are essential ideas in classical physics that define the link between the motion of an item and the forces acting upon it. These principles were proposed by Sir Isaac Newton in 1687 and remain crucial in comprehending a broad variety of physical events.

The first law, frequently referred to as the law of inertia, holds that an object at rest will stay at rest, and an object in motion will continue in uniform motion until acted upon by an external force. This indicates that things prefer to preserve their existing state of motion, whether it be at rest or in a straight line at a constant velocity. This rule offers a critical foundation for studying the behavior of things in the absence of forces.

The second rule of motion claims that the acceleration of an object is directly proportional to the net force applied on it and inversely proportional to its mass. This may be quantitatively represented as $F = ma$, where F represents the force exerted, m specifies the mass of the object, and a marks the consequent acceleration. This equation provides the quantitative link between force, mass, and acceleration, acting as a basic concept in physics and engineering.

The third rule, frequently phrased as "For every action, there is an equal and opposite response," highlights the reciprocal nature of forces. It says that if one thing exerts a force on a second object, the second object concurrently exerts an equal and opposite force on the first. This rule is fundamental in understanding the interactions between things and is especially relevant in the study of mechanical systems.

Together, these rules offer a complete framework for analyzing and predicting the behavior of physical things under varied contexts. They constitute the foundation for classical mechanics and are crucial in domains ranging from engineering to astronomy. Moreover, Newton's laws have not only survived the test of time but also led the way for the creation of more sophisticated theories, such as Einstein's theory of relativity, which expand and deepen our understanding of the world at both macroscopic and microscopic sizes.

2.2 Universal Law of Gravitation

The Universal Law of Gravitation, developed by Sir Isaac Newton in 1687, remains as one of the pillars of contemporary physics. This equation embodies the basic principle that every particle in the cosmos attracts every other particle with a force equal to the product of their masses and inversely proportional to the square of the distance between them. Newton's Law of Gravitation not only gave a precise mathematical description of how things interact owing to gravity, but it also supplied a coherent explanation for events seen both on Earth and in the skies.

This rule dramatically transformed humanity's perspective of the universe. Prior to Newton, theories for celestial movements were generally bogged in intricate, epicyclic models. The Universal Law of Gravitation effectively streamlined these explanations. It established that the same force guiding the fall of an apple from a tree was also responsible for the motion of the planets in their orbits around the Sun. This combination of terrestrial and celestial physics signified a paradigm change in scientific understanding.

The mathematical formulation of Newton's Law is attractive in its simplicity. The force of gravity (F) between two objects is provided by the product of their masses (m1 and m2), divided by the square of the distance (r) separating them, and multiplied by the gravitational constant (G). This equation, $F = G * (m1 * m2) / r^2$, not only permits exact calculations for forces on Earth but also offers the framework for forecasting the orbits of planets, moons, and celestial bodies across the universe.

Furthermore, Newton's Law of Gravitation set the path for major scientific discoveries. It played a vital part in the successful forecasts of cosmic phenomena, such as the return of Halley's Comet. Additionally, it permitted the creation of Kepler's Laws of Planetary Motion, giving a theoretical grounding for the actual discoveries made by Johannes Kepler.

However, it is vital to realize that Newton's Law of Gravitation, although exceptionally precise for most practical uses, was surpassed by Albert Einstein's General Theory of Relativity in the early 20th century. Einstein's theory gave a more sophisticated and thorough explanation of gravity, particularly in cases involving very huge or fast-moving objects.

Nevertheless, Newton's Law remains a crucial and accessible tool in a broad variety of applications, and its legacy continues in both astrophysical and engineering settings. In essence, the Universal Law of Gravitation serves as a monument to the strength of human mind and remains a lasting cornerstone of contemporary physics.

2.3 Kepler's Laws of Planetary Motion

Kepler's Laws of Planetary Motion are a core set of laws in celestial mechanics that explain the motion of planets in our solar system. Formulated by the German astronomer Johannes Kepler in the early 17th century, these rules radically transformed the way we understand planetary orbits and constituted a vital step toward the creation of modern astronomy and astrodynamics.

The first rule, called as the rule of Ellipses, asserts that planets revolve around the Sun in elliptical orbits, with the Sun placed at one of the two foci. This was a dramatic shift from the traditional geocentric concept, which maintained that heavenly bodies traveled in complete circles around Earth. Kepler's finding that orbits were truly elliptical altered our knowledge of planetary motion, offering a more precise and predictive model.

Kepler's second rule, the rule of Equal Areas, focuses on the speed at which a planet travels along its orbit. It argues that an imaginary line linking a planet to the Sun sweeps away equal regions in equal durations of time. This implies that a planet moves quicker in its orbit when it is closer to the Sun (perihelion) and slower when it is further away (aphelion). This rule nicely explains why planets do not travel at a constant pace throughout their orbits.

The third rule, known as the rule of Harmonies or Kepler's Harmonic Law, gives a mathematical link between the orbital period of a planet and its distance from the Sun. Specifically, it asserts that the square of a planet's orbital period is proportionate to the cube of its semi-major axis (the average distance from the Sun). This rule played a vital influence in the eventual development of Isaac Newton's law of universal gravitation.

Kepler's Laws heralded a paradigm change in astronomy, since they offered empirical support for the heliocentric model established by Nicolaus Copernicus. By rejecting the circular orbits and epicycles of prior models, Kepler's laws cleared the way for Isaac Newton's universal law of gravity. Newton's work, in turn, offered the theoretical basis to explain not just Kepler's laws but also a vast spectrum of astronomical occurrences.

Moreover, Kepler's Laws continue to have major practical implications in current space exploration. They serve as a basis for planning missions, calculating trajectories, and forecasting the locations of planets and spacecraft. The precision and accuracy of these principles have been proved by innumerable space missions, from interplanetary probes to the exact placing of communication satellites.

In conclusion, Kepler's Laws of Planetary Motion mark a major milestone in the history of astronomy and physics. They dramatically revolutionized our knowledge of planetary orbits, contradicting old models and providing a firm basis for the heliocentric perspective of our solar system. Beyond their historical relevance, these laws continue to play a crucial role in current astrodynamics and space exploration, highlighting the continuing impact of Johannes

Kepler's breakthrough contributions to celestial mechanics.

Chapter 3: Orbital Elements and Parameters

3.1 Semi-Major Axis, Eccentricity, and Inclination

"Semi-Major Axis, Eccentricity, and Inclination" are key principles in celestial mechanics that constitute the cornerstone of understanding and forecasting the motion of celestial bodies within a gravitational system. Each parameter gives crucial information about an object's orbit, giving insights into its shape, direction, and distance from the central body.

The semi-major axis, indicated by 'a', is a crucial quantity dictating the size of an orbit. It indicates half of the longest diameter of the elliptical route followed by a celestial body. In essence, it acts as a measure of the average distance of the item from the main body, whether it a planet, moon, or star. For instance, in the case of a planet circling the Sun, the semi-major axis defines its average distance from the Sun. This value is crucial in estimating orbital periods and knowing the energy of the orbiting body.

Eccentricity, commonly denoted by the symbol 'e', defines the form of an orbit. It measures how flattened or extended an orbit is relative to a perfect circle. Ranging from 0 (a circular orbit) to 1 (a parabolic trajectory), eccentricity defines the degree to which an orbit deviates from circularity. A high eccentricity suggests a more extended orbit, whereas a low one denotes a nearly circular course. For instance, comets, with their very eccentric orbits, display large fluctuations in distance from the Sun during their cycle.

Inclination, represented by 'i', applies to the tilt of an orbit relative to a reference plane, commonly regarded as the plane of the main body's equator. It determines the angle between the orbital plane and the reference plane. A zero inclination denotes an orbit being in the same plane as the reference, while a 90-degree inclination results in a polar orbit. For instance, the inclination of Earth's orbit around the Sun is around 23.5 degrees, leading to the shifting seasons.

These three parameters, semi-major axis, eccentricity, and inclination, combined offer a thorough description of an orbit's features. By comprehending these factors, astronomers and astrodynamists may precisely anticipate an object's passage through space, calculate orbital periods, and organize interplanetary missions. Additionally, these characteristics are vital for constructing spacecraft trajectories and for analyzing the dynamics of celestial systems, making them indispensable instruments in the area of celestial mechanics.

3.2 Argument of Periapsis, True Anomaly, and Mean Anomaly

In celestial mechanics, knowing the complexities of orbital dynamics is vital for proper space mission design and analysis. Three critical metrics, the Argument of Periapsis, True Anomaly, and Mean Anomaly, play crucial roles in characterizing and forecasting an object's location inside its orbit.

The Argument of Periapsis is a basic quantity that governs the direction of an orbit inside a plane. Specifically, it defines the angle between the ascending node, the point at which an orbit crosses a reference plane from below, and the periapsis, the closest point of approach to the main body. This parameter carries crucial consequences for missions requiring spacecraft rendezvous, since it alters the relative locations of objects at various points in their orbits. For instance, a spacecraft's Argument of Periapsis may be intentionally altered to maximize rendezvous maneuvers with a target celestial body.

On the other hand, the True Anomaly encompasses the present location of an item inside its orbit. It quantifies the angle between the periapsis, the closest approach to the primary body, the object's present location, and the primary body. This parameter gives real-time information on an object's location in its orbit, allowing for exact estimates of position, velocity, and required orbital changes. For example, in interplanetary missions, calculating the True Anomaly is critical for designing maneuvers like as orbital insertions or gravity assistance.

The Mean Anomaly, in contrast, represents the average angular distance of an item along an elliptical orbit for a particular time. It is a time-dependent parameter that gives an analytical tool for estimating future locations inside an orbit. The Mean Anomaly serves as a reference point for estimating locations at specified epochs, enabling mission planning for rendezvous, intercepts, and other orbital activities. By exploiting mathematical correlations between the

Mean Anomaly and time, engineers may properly estimate orbital locations long into the future.

These three characteristics operate in combination to offer a thorough knowledge of an object's orbital behavior. The Argument of Periapsis establishes the direction of an orbit, the True Anomaly offers the immediate position, and the Mean Anomaly provides a time-dependent estimate of an object's location. When combined, these characteristics constitute a strong toolset for mission planners and astrodynamists, allowing them to accurately direct spacecraft through the challenges of space. Furthermore, developments in numerical integration methods have boosted the accuracy of forecasts incorporating these parameters, revolutionizing space exploration and satellite operations. Overall, a detailed grasp of these orbital aspects is important for effective and efficient navigation throughout the universe.

3.3 State Vectors and Six Orbital Elements

"State Vectors and Six Orbital Elements" are essential ideas within celestial mechanics, giving a robust framework for describing and forecasting the motion of celestial bodies. State vectors and orbital components give diverse but complementary viewpoints on the mobility of objects in space.

State vectors are a collection of mathematical values that completely define the location and velocity of an item at a certain instant in time. This format is especially beneficial for numerical simulations and real-time monitoring of celestial bodies. A state vector generally consists of three location coordinates (x, y, z) and three velocity components (vx, vy, vz). By incorporating both location and velocity information, state vectors allow exact forecasts of future positions and velocities, as well as retrodictions of previous states.

In comparison, the six orbital components give a more natural and geometric explanation of an orbit. These elements include the semi-major axis (a), eccentricity (e), inclination (i), argument of periapsis (ω), longitude of the ascending node (Ω), and true anomaly (v). The semi-major axis controls the size of the orbit, while eccentricity specifies its form, ranging from circular (e=0) to extremely elliptical (e<1) and parabolic (e=1) to hyperbolic (e>1). Inclination specifies the tilt of the orbit relative to a reference plane, whereas the argument of periapsis and longitude of the ascending node indicate the orientation of the orbit within that plane. The actual anomaly reveals the location of the object along its orbit at any given moment.

These two representations, state vectors and orbital elements, are coupled by mathematical transformations. Converting between them offers for flexibility in studying and addressing issues in celestial mechanics. For instance, state vectors are commonly employed in numerical simulations, where accurate numerical integration methods are utilized to forecast future locations. Conversely, orbital components are useful for mission planning and design, allowing a more intuitive knowledge of the orbit's features.

Understanding the link between state vectors and orbital components is vital for numerous applications in space exploration and satellite operations. For example, while planning interplanetary missions, state vectors are applied for exact maneuver planning and execution, assuring accurate rendezvous and insertion into targeted orbits. On the other hand, the six orbital components give crucial insights for the selection of appropriate orbits for particular missions, such as geostationary orbits for communication satellites or heliocentric orbits for planetary missions.

In conclusion, "State Vectors and Six Orbital Elements" constitute the cornerstone of celestial mechanics, offering unique but complimentary accounts of the motion of celestial entities. State vectors enable exact numerical representations of location and velocity, permitting real-time monitoring and simulations. Meanwhile, the six orbital components give a geometric knowledge of orbits, assisting in mission planning and design. The interaction between these two notions is crucial for a thorough understanding of the dynamics of objects in space and is vital to the success of space exploration operations.

Chapter 4: Two-Body Problem and Keplerian Orbits

4.1 Kepler's Equation and Solving for Anomalies

"Kepler's Equation and Solving for Anomalies" is a key subject in celestial mechanics, crucial for understanding the motion of celestial bodies in elliptical orbits. Johannes Kepler's breakthrough work in the early 17th century provided the framework for contemporary astrodynamics, notably in explaining the motion of planets around the Sun. Kepler's Equation originates from his first rule, which asserts that planets travel in elliptical orbits with the Sun at one of the foci.

Kepler's Equation may be stated as $M = E - e \sin(E)$, where M is the mean anomaly, E is the eccentric anomaly, and e is the eccentricity of the orbit. The mean anomaly is an angular value that grows linearly with time and defines the location of a celestial body during its elliptical orbit. On the other side, the eccentric anomaly refers to the angle between the semi-major axis and the position vector of the body in its orbit.

Solving Kepler's Equation is a non-trivial problem owing to its transcendental character. Historically, mathematicians and astronomers have devised numerous numerical algorithms and iterative procedures to discover solutions. One typical strategy is the Newton-Raphson method, which repeatedly refines an initial approximation for E based on the functional form of Kepler's Equation. This iterative technique converges swiftly and produces precise results for E.

Understanding anomalies is vital for forecasting the location of celestial bodies properly. The eccentric anomaly, in instance, is an intermediate measure that helps define the real anomaly—the angle between the periapsis and the present location of the body. This, in turn, assists in forecasting the location of celestial objects at any given moment, which is vital for spacecraft navigation and mission planning.

Kepler's Equation also offers the foundation for calculating the time of flight and orbital period, both of which are crucial elements in mission planning. Engineers and astrodynamists depend on these calculations to determine trajectories for spacecraft, ensuring they reach their destinations with accuracy.

Moreover, Kepler's Equation has played a vital role in the research of exoplanets. Observations of exoplanetary transits and radial velocity fluctuations are evaluated using Kepler's Equation to deduce essential aspects of these faraway planets, such as their orbital eccentricity, semi-major axis, and even their atmospheres.

In essence, "Kepler's Equation and Solving for Anomalies" constitutes a cornerstone in celestial mechanics. It gives a mathematical foundation for studying the motion of celestial entities in elliptical orbits. The iterative techniques established to solve this equation have practical implications in space mission planning, navigation, and the investigation of exoplanets. Without a question, this subject serves as a tribute to the ongoing influence of Johannes Kepler's revolutionary work in the realm of astrodynamics.

4.2 Classical Orbital Elements

Classical Orbital Elements are a key idea in celestial mechanics, giving a clear and powerful approach to characterize the state of an orbiting object inside a gravitational system. These components determine the form, size, and direction of an orbit, allowing for exact prediction and study of its movements.

The first element, the semi-major axis (a), is a key parameter. It represents the average distance between the orbiting body and the main body, and effectively determines the size of the orbit. This number is a key feature, since it directly corresponds to the period of the orbit via Kepler's third law. A bigger semi-major axis denotes a broader orbit, whereas a lower number signifies a more compact trajectory.

Eccentricity (e) is another crucial aspect. It measures the divergence of the orbit from a complete circle. An eccentricity of 0 denotes a circular orbit, whereas higher values imply an elliptical one. This characteristic determines the pace at which the orbiting body travels along its path; larger eccentricities lead to more substantial fluctuations in speed.

Inclination (i) denotes the tilt of the orbital plane relative to a reference plane. It's a critical factor, since it specifies the orientation of the orbit in three-dimensional space. Inclination is vital for understanding the interaction between multiple celestial entities within a multi-body system and for planning interplanetary missions.

The argument of periapsis (ω) is the angle between the ascending node (the point at which the orbit crosses the reference plane from below) and the periapsis (the point of closest approach to the main body). This element offers information on the orientation of the orbit inside the orbital plane. It helps specify the particular route that the orbit takes as it travels its ellipse.

True anomaly (v) completes the set of classical orbital elements. It represents the current location of the orbiting body along its trajectory, measured from the periapsis. This element delivers a real-time measure of the location of the object in its orbit, allowing for exact forecasts of its future position.

Collectively, classical orbital components give a simple and strong depiction of an orbit's state. They allow exact estimates of future locations, which are vital for mission planning, satellite tracking, and understanding celestial dynamics. Additionally, these parts provide the foundation for more advanced studies, such as perturbation theory, which examines external effects on an orbit. Overall, classical orbital components are basic in the subject of celestial mechanics, giving the basis for a broad variety of applications in space exploration and astrodynamics.

4.3 Special Orbits: Circular, Elliptical, Parabolic, Hyperbolic

Celestial mechanics involves a vast variety of orbits, each displaying different traits and behaviors. Among them, four essential kinds stand out: Circular, Elliptical, Parabolic, and Hyperbolic orbits. These orbits serve key roles in understanding and planning space missions, each presenting distinct benefits and problems.

Circular orbits are delightfully basic in their geometry, possessing a constant radius around the center body. They are characterized by a single constant angular velocity, making them especially effective for satellites in applications needing stable locations relative to Earth, such as communication satellites in geostationary orbits. The balance of forces in a circular orbit results in a permanent motion that doesn't need any more changes, making it a desirable option for long-term missions.

Elliptical orbits, on the other hand, give a greater variety of distances from the central body. These orbits are distinguished by their eccentricity, which defines how 'stretched out' or 'squished' the elliptical becomes. Highly elliptical orbits, such as Molniya orbits, are particularly helpful for satellites needing lengthy stay durations over certain portions of the Earth, including those used for communication in polar regions. Elliptical orbits are very prevalent in interplanetary missions, enabling spacecraft to easily travel huge distances while altering their speed at periapsis and apoapsis.

Parabolic orbits describe a specific class of paths where the orbital eccentricity approaches unity. This indicates that the object's velocity is exactly the escape velocity at the time of closest approach. Parabolic orbits are fundamentally open, suggesting that they don't shut on themselves like circular or elliptical orbits. Objects on parabolic trajectories are typically related with celestial bodies on hyperbolic trajectories, such as comets traveling through the solar system.

Hyperbolic orbits, in contrast, indicate routes where an object's velocity surpasses the escape velocity at any point along its trajectory. These orbits are free and unbounded, meaning that the item will ultimately escape the gravitational attraction of the central body. Hyperbolic paths are usually encountered in astronomical occurrences such as the flybys of interstellar objects like 'Oumuamua, which provide unique chances for scientific inquiry.

In practical terms, the choice of orbit type is critical for mission planning and execution. Circular orbits are appropriate for long-term station-keeping missions, whereas elliptical orbits give flexibility in coverage and mission length. Parabolic and hyperbolic orbits, while less frequent for artificial satellites, play an important role in the study of celestial bodies and in rare interstellar contacts. Understanding and mastering these particular orbits are crucial abilities for astrodynamics engineers and space scientists alike, allowing them to traverse and explore the universe with precision and purpose.

Chapter 5: Perturbations and Non-Keplerian Orbits

5.1 Perturbation Theory

Perturbation Theory is a strong mathematical technique applied in the area of celestial mechanics to explore the impact of minor disturbances on the motion of celestial bodies inside a gravitational system. It gives a technique to estimate the trajectory of a perturbed orbit by breaking down the issue into smaller, solvable components. This theory is especially relevant when dealing with real-world settings when gravitational forces from many bodies or other external factors differ from the idealized two-body issue.

One fundamental component of Perturbation Theory rests in its capacity to formulate the equations of motion as a sum of terms, each reflecting a distinct level of perturbation. These variables may then be studied separately, providing for a step-by-step comprehension of how distinct forces contribute to the total motion. This decomposition gives a methodical way to understanding complicated systems, making it a vital tool for space missions, satellite navigation, and the investigation of celestial occurrences.

In actuality, Perturbation Theory is particularly essential when dealing with objects in Earth's orbit, since the existence of other celestial bodies, such as the Moon or other planets, brings extra gravitational forces. By employing Perturbation Theory, astrodynamists may produce exact predictions regarding satellite locations, allowing accurate orbit determination and aiding operations like rendezvous, docking, and space debris abatement. This theory also underlies the correct computation of interplanetary trajectories, vital for planning expeditions to distant celestial planets.

Furthermore, Perturbation Theory is not restricted to gravitational interactions alone. It may be expanded to account for additional perturbing factors including air drag, solar radiation pressure, and magnetic interactions. This adaptability enables it to be used to a broad variety of situations, including the planning and execution of space missions, the research of long-term stability in celestial systems, and the investigation of space debris dispersion.

Despite its potency and applicability, Perturbation Theory does have its limits. It presupposes that perturbations are modest, and higher-order terms are generally overlooked in reality. In excessively dynamic or chaotic systems, these simplifying assumptions may lead to less accurate forecasts. Moreover, Perturbation Theory may not be appropriate in circumstances when relativistic effects become considerable, requiring more sophisticated formulations utilizing Einstein's General Theory of Relativity.

In conclusion, Perturbation Theory serves a significant role in celestial mechanics by giving an organized way to studying the effects of minor disturbances on the motion of celestial bodies. Its capacity to break down complicated systems into manageable components is crucial for producing exact forecasts in situations ranging from satellite navigation to interplanetary mission planning. However, it's crucial to be cognizant of its limits, particularly in circumstances when higher-order perturbations or relativistic effects come into play. Overall, Perturbation Theory serves as a cornerstone in the toolset of astrodynamists and space mission planners, allowing the effective navigation and exploration of the universe.

5.2 Mean Elements and Osculating Elements

"Mean Elements and Osculating Elements" are fundamental ideas in the science of celestial mechanics, providing astronomers and astrodynamists with critical tools for precisely forecasting and evaluating the motion of celestial bodies. These factors are crucial in reducing complicated, time-varying orbits into more manageable parameters.

Mean Elements offer a smoothed, averaged description of an orbit. They give a robust and regularized framework for defining the motion of celestial bodies. By reducing short-term disturbances and abnormalities, mean elements allow astrodynamists concentrate on the long-term behavior of the orbit. This simplification is especially beneficial when working with multi-body systems, because interactions among numerous gravitational effects may make direct study of the orbit problematic.

On the other hand, Osculating Elements reflect the instantaneous features of an orbit at a certain moment in time. They capture the real state of the orbit, including any perturbations and fluctuations owing to external factors. Osculating elements are dynamic and may change considerably over short times, reflecting the genuine, time-varying character of the orbit. This makes them important for activities like mission planning and maneuver execution, since they offer exact information regarding the spacecraft's position and velocity at any given instant.

Mean Elements and Osculating Elements are related by a technique known as "averaging theory." This mathematical approach includes integrating the equations of motion over a specific time period to derive mean elements from osculating components. This transformation allows for the simplification of complicated, disturbed orbits into a more stable, averaged form. By switching between mean and osculating components, astrodynamists may pick the most suited representation for a certain job or investigation.

The option between utilizing mean elements or osculating elements relies on the context of the task at hand. For long-term trajectory planning, such as interplanetary missions, mean components are favored owing to their stability and regularity. They enable for accurate forecasts over long periods, disregarding short-term fluctuations. Conversely, for operations needing real-time accuracy, such spacecraft rendezvous or orbital insertion maneuvers, osculating components are required. They give the most up-to-date information regarding the orbit, allowing accurate navigation and control.

In summary, Mean Elements and Osculating Elements are crucial tools in celestial mechanics, presenting alternative viewpoints on the motion of celestial bodies. Mean Elements give a stable, averaged representation of an orbit, appropriate for long-term predictions and studies. Osculating Elements, on the other hand, give an immediate perspective of the orbit, incorporating all perturbations and variations at a single point. The capacity to move between these representations provides astrodynamists the flexibility required to face a broad variety of issues in space exploration and navigation.

5.3 J2 Perturbation and Geopotential Models

Celestial mechanics involves a myriad of elements that impact the motion of things in space. Among them, J2 perturbation and geopotential models stand out as key factors in understanding the dynamics of Earth's satellites and spacecraft. The J2 perturbation, also known as the oblateness disturbance, derives from the Earth's non-spherical form. This divergence from a perfect sphere causes gravitational fluctuations, notably noticeable in orbits with greater altitudes. The second zonal harmonic, abbreviated as J2, quantifies this oblateness impact, and it plays a crucial role in perturbing satellite trajectories.

Geopotential models give a mathematical picture of Earth's gravitational field, allowing exact estimates of satellite orbits. These models incorporate numerous harmonics beyond the second order, allowing for progressively detailed computations. The use of geopotential models is crucial in current astrodynamics, allowing exact orbit determination and prediction. Notably, J2 perturbation is one of the most critical factors in these models, influencing orbits in ways that cannot be disregarded, particularly in applications where high accuracy is required.

The influence of J2 perturbation becomes noticeable in satellite orbits at medium to high altitudes, when departures from perfectly Keplerian motion become significant. As a satellite circles the Earth, it encounters fluctuations in the gravitational pull, resulting to perturbations in its course. Over time, these disturbances compound, possibly leading to considerable disparities between the projected and real orbits. Geopotential models, which integrate J2 and higher-order components, give a mechanism to account for these effects, allowing for reliable forecasts of satellite locations over lengthy periods.

One of the primary repercussions of J2 disruption is the precession of the satellite's orbit. This precession originates owing to the oblate form of the Earth, causing the orientation of the orbital plane to steadily alter over time. This phenomena has practical significance for satellite operators and mission planners. Understanding and managing the impacts of J2 perturbation is vital for maintaining the required operating orbits, especially for satellites with critical missions like as Earth observation or global navigation systems.

Geopotential models are regularly developed and updated to account for more gravitational harmonics and to attain better accuracy in orbit determination. Advanced models, such as the EGM (Earth Gravitational Model) series, integrate a plethora of harmonics beyond J2, offering an extraordinarily realistic depiction of Earth's gravitational field. These models find uses not just in satellite navigation and tracking but also in sectors like geodesy and sea level monitoring.

In summary, the impact of J2 perturbation and geopotential models in celestial mechanics cannot be emphasized. The non-spherical form of the Earth creates departures from traditional Keplerian motion, particularly for satellites in medium to high altitudes. J2 disruption leads to problems like orbital precession, which have practical ramifications for satellite operations. Geopotential models, comprising J2 and higher-order terms, are crucial in reducing these effects and assuring correct orbit determination. As technology progresses, these models continue to change, offering even more accurate navigation and tracking capabilities in Earth's dynamic environment.

Chapter 6: Three-Body and n-Body Problems

6.1 The Restricted Three-Body Problem

The Restricted Three-entity Problem is a key topic in celestial mechanics that gives a simpler framework for understanding the motion of a smaller entity under the influence of two bigger, gravitational masses. In this scenario, one examines a massless object, frequently referred to as the "test particle," interacting with two substantial celestial bodies, often a star and a planet. The impact of the test particle on the bigger bodies is overlooked, thereby presuming that the test particle's mass is small compared to the other two. This reduction allows for a more manageable mathematical description of the system.

The Restricted Three-Body Problem was initially proposed in the 18th century by Euler and Lagrange, but its contemporary form and extensive analysis were greatly developed by the mathematician Henri Poincaré in the late 19th and early 20th centuries. Poincaré's work established the framework for a fuller understanding of the complex and frequently unexpected behavior displayed by such systems. He proved that even within this reduced framework, chaotic behavior may develop under certain circumstances, leading to the discipline of dynamical systems theory and chaos theory.

One of the important properties of the Restricted Three-Body Problem is the occurrence of Lagrange points, which are five precise positions in the system where the gravitational pulls of the two bigger bodies and the centrifugal force of the spinning frame balance exactly. These ideas have substantial significance for space missions and the stability of celestial bodies. For instance, Lagrange points L1, L2, and L3 sit along the line linking the two bigger bodies and are termed unstable equilibrium points. Lagrange points L4 and L5, on the other hand, form the vertices of an equilateral triangle with the two bigger bodies and are stable. These sites have been of tremendous interest for missions like the James Webb Space Telescope and several Earth-observing satellites.

Furthermore, the Restricted Three-Body Problem offers a framework for understanding resonance phenomena in celestial mechanics. For example, the Trojan asteroids associated with Jupiter are in a 1:1 mean-motion resonance with the gas giant, meaning they circle the Sun at the same average rate as Jupiter, and are positioned between its L4 and L5 locations. This resonance results in stable orbits, and comparable resonances may be found in other astronomical systems, contributing to the stability and structure of planetary rings, moon systems, and other complicated orbital configurations.

In practical terms, the Restricted Three-Body Problem has found applications in the construction of trajectories for space missions. By leveraging the dynamics of Lagrange points and resonances, spacecraft may be positioned in stable orbits with little energy consumption. For instance, the employment of gravity aids, which use the gravitational pull of celestial entities to adjust a spacecraft's course, is a direct consequence of the ideas outlined by the Restricted Three-Body Problem.

In conclusion, the Restricted Three-Body Problem is a significant theoretical framework in celestial mechanics, allowing for the investigation of intricate interactions between three enormous masses in space. Its creation and study have led to fundamental discoveries into the behavior of celestial systems, including the finding of Lagrange points, the comprehension of resonance phenomena, and the planning of efficient space missions. This problem's lasting relevance is seen in its continuous applicability in current astrodynamics and space exploration.

6.2 Lagrange Points and Halo Orbits

Lagrange Points, also known as libration points, are five unique regions in space where the gravitational pulls of two big entities, such as the Earth and the Moon or the Earth and the Sun, balance the centripetal force felt by a smaller object, such a spacecraft. These points were found by the French mathematician Joseph-Louis Lagrange in 1772. They have gained substantial interest in celestial mechanics owing to their remarkable stability qualities.

The first three Lagrange Points, L1, L2, and L3, sit along the line linking the two bigger bodies. L1, known as the "Lagrangian Point 1," is especially essential since it maintains a consistent alignment with the Earth and the Sun. This makes it a perfect site for numerous space observatories and spacecraft, such as the Solar and Heliospheric Observatory (SOHO) and the James Webb Space Telescope (JWST). These observatories can acquire critical data about the Sun and distant astronomical objects without being blocked by the Earth's atmosphere.

L2, on the other hand, is positioned on the opposite side of the Earth from the Sun. While L2 is not as widely used as L1, it has been exploited for missions like the Wilkinson Microwave Anisotropy Probe (WMAP) and the European Space Agency's (ESA) Herschel Space Observatory, because to its unimpeded vision of deep space.

L3, while theoretically stable, is seldom utilized in reality. It resides on the opposite side of the Sun from the Earth and is regarded less tactically beneficial for most missions.

Beyond the three collinear Lagrange Points, there are two more Lagrange Points, L4 and L5, which form an equilateral triangle with the bigger bodies. These are commonly referred to as the Trojan Points. Objects at these positions tend to revolve around the Lagrange Point rather than the two bigger bodies, producing stable orbits. This behavior has been noticed in the Trojan asteroids that follow Jupiter in its orbit around the Sun.

Halo orbits, often termed Lissajous orbits, are a special sort of three-dimensional periodic orbit around Lagrange Points. These orbits enable spacecraft to "hover" over a Lagrange Point, retaining a relatively stable location with regard to the two bigger planets. Halo orbits are exceedingly stable, making them excellent for long-term missions such as space observatories, since they need minimum fuel for station-keeping.

In summary, Lagrange Points and Halo Orbits play a key role in current space exploration. They give stable places in orbit where spacecraft may function successfully, making them crucial for missions like space observatories and exploration expeditions to distant celestial bodies. Understanding and exploiting these points have changed our abilities to study and explore the universe.

6.3 Generalized n-Body Problems

The Generalized n-Body Problem provides a considerable expansion of the standard two-body and three-body problems, involving situations with more than three interacting celestial bodies. Unlike the limited three-body issue, which includes one dominant body and two lesser ones, the generalized n-body problem doesn't make any assumptions on mass disparities or particular force interactions. This makes it a very sophisticated and flexible topic of research within celestial mechanics.

One of the primary obstacles in addressing the generalized n-body issue consists in the large amount of interactions between many bodies. Each body exerts gravitational pressures on every other, resulting in a complicated web of relationships. The resultant equations of motion create a set of linked, non-linear differential equations, which are notoriously difficult to solve analytically. Consequently, numerical approaches and computer simulations frequently become key tools in studying complex systems.

Moreover, the generalized n-body issue is vital for understanding the dynamics of celestial systems like star clusters, galaxies, and other cosmic structures where numerous things interact gravitationally. It also plays a key role in the study of planetary systems, especially those displaying complicated resonances or interactions between multiple moons or asteroids.

This complexity may rise to phenomena like chaotic behavior, where apparently modest perturbations can lead to dramatically divergent results over time.

A crucial component of the generalized n-body issue is its importance in astrodynamics and space missions. For example, mission planners and astrodynamists must direct spacecraft across areas of space where several celestial bodies exert gravitational forces. These missions need accurate trajectory planning and optimization to accomplish desired results, such as rendezvous with celestial bodies or the optimal use of gravitational aids for trajectory modifications.

In addressing the generalized n-body issue, researchers generally resort to sophisticated numerical integration methods. These include well-known techniques like the Runge-Kutta algorithm and more advanced approaches such as symplectic integrators, which especially suitable for saving energy and angular momentum during extended integrations. Such approaches allow the study of complicated, multi-body interactions over longer periods of time, offering insights into the stability and long-term behavior of these systems.

In conclusion, the Generalized n-Body Problem provides a key field of research within celestial mechanics, expanding beyond the basic two-body and three-body situations. It gives insight into the complicated interactions between several celestial entities, with applications ranging from comprehending cosmic structures to permitting accurate space missions. The numerical approaches applied in tackling these issues play a crucial role in understanding the complicated dynamics of these multi-body systems. As our grasp of celestial physics continues to grow, so too will our capacity to navigate and comprehend the intricacies of the universe.

Chapter 7: Orbital Maneuvers and Hohmann Transfers

7.1 Impulsive and Continuous Thrust Maneuvers

"Impulsive and Continuous Thrust Maneuvers" are two main ways to modifying the trajectory of spacecraft in celestial physics. These movements play a key role in the conduct of space missions, impacting everything from satellite placement to interplanetary flight.

Impulsive movements include a quick and immediate shift in velocity. This shift is often done by the firing of onboard propulsion systems, such as rockets or thrusters, for a limited time. These maneuvers are distinguished by their high thrust-to-weight ratio and are well-suited for activities like orbital transfers or rendezvous operations. The mathematical description of impulsive movements simplifies the computations, allowing for exact predictions of trajectory modifications. However, impulsive movements are susceptible to limits in terms of energy economy and their applicability for long-duration missions.

On the other hand, continuous thrust movements involve a sustained application of force over a long duration. This strategy depends on engines that can function for longer periods, such as ion or electric propulsion systems. Continuous thrust gives advantages in terms of fuel economy, allowing spacecraft to attain high velocities while spending less propellant compared to classical chemical propulsion. This makes constant thrust movements vital for distant space missions, such as those involving outer planets or interstellar travel. However, the complexity of predicting fluctuating thrust levels and non-uniform accelerations provide distinct obstacles in mission planning.

The decision between impulsive and continuous thrust operations relies on mission-specific needs and goals. For instance, impulsive maneuvers are chosen for missions when quick velocity changes are essential, such as either entering or exiting planetary orbits. Conversely, continuous thrust maneuvers are better suited for missions that need efficiency and lengthy operating durations, making them excellent for missions to distant celestial bodies or for spacecraft requiring station-keeping in geostationary orbits.

In actuality, a mix of both maneuver types is typically deployed in complicated missions. For instance, a spacecraft may utilize impulsive maneuvers to execute substantial trajectory modifications, while applying continuous push for fine-tuning its orbit or completing lengthy missions in deep space. This hybrid technique exploits the characteristics of each maneuver type, improving both accuracy and efficiency in space missions.

In summary, impulsive and continuous thrust maneuvers are two key strategies in celestial mechanics for modifying the trajectory of spacecraft. Impulsive motions allow quick, instantaneous changes in velocity, making them suited for activities needing urgent corrections. Continuous thrust maneuvers, on the other hand, depend on continuous propulsion over a long time, emphasizing fuel economy and permitting trips to distant celestial planets. The decision between these maneuvers relies on mission-specific goals and needs, typically resulting to a hybrid of both tactics for effective mission performance.

7.2 Hohmann Transfer Orbits

Hohmann Transfer Orbits are a key concept in celestial mechanics, providing an effective technique of moving a spacecraft from one circular orbit to another within the same gravitational field, such as between planets or moons. Proposed by Walter Hohmann in 1925, this maneuver takes use of the conservation of energy concept in an ideal two-body system. It includes two impulses: the first to increase velocity at the beginning orbit, and the second to reduce velocity at the destination orbit.

The primary element of a Hohmann Transfer is its efficiency, lowering the necessary energy for the transfer. This is vital in space travel, as energy saving plays a key part in mission planning. By exploiting the concepts of energy conservation and the Oberth effect (where increasing speed in a gravitational field takes less energy for an extra velocity shift), Hohmann transfers provide an easy and inexpensive alternative for interplanetary missions.

One of the most noteworthy uses of Hohmann Transfer Orbits was in the early interplanetary missions of the 1960s and 70s. For example, the Mariner missions, including the momentous Mariner 9 which orbited Mars in 1971, required Hohmann transfers to reach their respective destinations. Such transfers considerably lowered the fuel needs compared to other trajectory possibilities, enabling for longer missions and the investigation of several celestial planets during a single mission.

However, it is crucial to note that although Hohmann transfers are very efficient, they are based on a simplified model that assumes just the gravitational pull of the main body (e.g., a planet) and ignores additional perturbing forces. In reality, celestial bodies are susceptible to the gravitational attraction of many objects, plus additional disturbances like as solar radiation pressure and atmospheric drag. As a consequence, mission planners frequently integrate extra maneuvers or corrections to account for these issues.

Moreover, Hohmann transfers are not always viable for all missions. For example, trips to planets with extremely inclined or eccentric orbits may need more complicated trajectory designs. Additionally, missions to comets or asteroids sometimes involve non-

Hohmann transfers because to the erratic structure of their orbits.

In conclusion, Hohmann Transfer Orbits have been significant in influencing the area of space exploration. Their efficiency and relatively basic implementation make them a cornerstone in mission planning. While they give a good beginning point for interplanetary travel, it's vital for mission planners to consider other aspects and possible modifications to achieve a successful and precise trajectory. The notion of the Hohmann Transfer remains a monument to the beauty of celestial mechanics and its practical applications in space exploration.

7.3 Bi-Elliptic Transfers and Interplanetary Trajectories

Bi-Elliptic Transfers and Interplanetary paths reflect sophisticated ideas in celestial physics, affording engineers and space mission planners significant tools for optimizing spacecraft paths inside our solar system.

A Bi-Elliptic Transfer is a two-burn orbital maneuver that permits a spacecraft to shift between two coplanar circular orbits of differing radii. The essential aspect of this transfer is its potential for great efficiency when the ratio of final to beginning orbit radius is large. This maneuver comprises an initial fire to lift the spacecraft into a highly elliptical orbit, followed by a second burn at apoapsis to circularize the orbit at the target radius. The trade-off of time for efficiency is considerable, making Bi-Elliptic Transfers especially effective for missions with flexible time requirements but stringent propulsion limits.

One of the primary benefits of the Bi-Elliptic Transfer is in its potential to achieve substantial changes in orbit without incurring excessive fuel consumption. This makes it a viable strategy for missions targeting distant celestial entities or attempting to maximize resource consumption. However, the negative of prolonged transfer periods might be a limiting issue for time-sensitive missions or those needing fast reactions to dynamic occurrences in space.

Interplanetary Trajectories, on the other hand, comprise a larger class of trajectories utilized to move between planets. These trajectories entail sophisticated computations, including parameters like as the locations and velocities of both the origin and destination planets, orbital transfers, and the timing of launches to take advantage of good planetary alignments. The Hohmann Transfer, for instance, is a regularly utilized interplanetary route that utilizes the idea of minimal energy transfer between two coplanar orbits.

Additionally, gravity assistance may be deployed to increase the efficacy of interplanetary paths. By leveraging a near approach of a celestial body, such as a planet or moon, spacecraft may acquire or lose energy, modifying their trajectory in a manner that is not achievable with normal rocket power alone. This approach has been vital in permitting trips to outer planets like Jupiter and Saturn, where large distances and limited propulsion capabilities face severe hurdles.

While Interplanetary Trajectories provide a flexible framework for journeys between planets, they need exact calculations and precise execution. Errors in course changes or orbital insertion may cause to mission failure or unsatisfactory outcomes. Therefore, the meticulous design and execution of interplanetary paths remain key components of successful space travel.

In summary, Bi-Elliptic Transfers and Interplanetary Trajectories constitute two key notions in celestial physics. Bi-Elliptic Transfers enable a trade-off between time and fuel economy, making them appropriate for missions with severe propulsion limits but permissive timelines. Meanwhile, Interplanetary Trajectories constitute the backbone of interplanetary missions, requiring complicated calculations and gravity aids to travel around the solar system with accuracy. Both strategies stand as witness to the amazing creativity and mathematical skill that underpin current space exploration.

Chapter 8: Rotational Dynamics and Attitude Control

8.1 Euler's Equations of Motion

"Euler's Equations of Motion" are a series of basic equations in celestial mechanics that explain the rotational motion of a rigid body. Named after the Swiss mathematician and physicist Leonhard Euler, these equations give a mathematical framework for understanding the motion of an object in three-dimensional space. They are significant in several industries, including aeronautical engineering, robotics, and physics.

At their heart, Euler's Equations define the connection between the external torques acting on a spinning body and the ensuing changes in its angular velocity. The equations take into consideration the body's moments of inertia, which indicate how mass is distributed inside the item. This implies that Euler's Equations are especially effective when dealing with objects with complicated forms or uneven mass distributions.

One of the fundamental properties of Euler's Equations is their generality. They apply to both spinning rigid bodies in classical mechanics and to quantum mechanical systems, proving their wide relevance across many disciplines of physics. This adaptability has made them a cornerstone in the investigation and modeling of diverse physical systems.

Euler's Equations consist of three connected, second-order differential equations, each corresponding to a distinct axis of rotation (x, y, and z). These equations show how the angular velocity vector varies with respect to time under the effect of external torques. As a consequence, they are especially significant in the design and control of spacecraft, satellites, and other rotating systems.

Furthermore, Euler's Equations illustrate the conservation of angular momentum, a key fact in physics. They demonstrate that in the absence of external torques, the angular momentum of a closed system stays constant. This concept is fundamental in the study of the behavior of celestial bodies and artificial satellites in space.

In practical applications, Euler's Equations find substantial usage in the creation of control systems for aeronautical vehicles. By controlling the external torques given to a spacecraft, engineers may accurately regulate its orientation and location in space. This is vital for mission success, since keeping the right orientation is typically essential for operations like communication, navigation, and data collecting.

In summary, Euler's Equations of Motion are a cornerstone of celestial mechanics, giving a strong tool for understanding the rotational dynamics of rigid bodies. Their generality and adaptability have led to numerous applications in domains ranging from aeronautical engineering to robotics. By explaining the link between torques, angular velocity, and moments of inertia, Euler's Equations allow exact control and knowledge of rotational motion in three-dimensional space. This makes them a vital tool in the design and operation of complex systems in the universe.

8.2 Attitude Representation: Euler Angles, Quaternions

Attitude Representation is an important part of celestial mechanics and spacecraft dynamics, permitting the exact representation of an object's orientation in space. Two typical approaches adopted for this purpose are Euler Angles and Quaternions.

Euler Angles, named after the Swiss mathematician Leonhard Euler, give an accessible method to depict the orientation of a body. They define the sequential revolutions around three independent axes, commonly referred to as pitch (rotation about the y-axis), roll (rotation about the x-axis), and yaw (rotation about the z-axis). Euler angles are intuitive and simple to depict, making them a popular option for many technical applications. However, they are not without restrictions. The issue of gimbal lock, whereby a specific orientation might lead to a loss of one degree of freedom, can be a major obstacle when utilizing Euler angles. This constraint may lead to computational instability in some instances, especially when the object executes complicated movements.

On the other hand, Quaternions provide an alternate and strong technique to expressing attitude. They are a kind of hypercomplex numbers having four components - a scalar portion and a vector part. This representation eliminates the problem of gimbal lock associated with Euler angles, since quaternions work in a three-dimensional space, unlike Euler angles which act in a two-dimensional environment. Quaternions are highly preferred in computational applications because to their numerical stability, making them a great option for simulations and real-time control systems. However, they might be more hard to comprehend intuitively compared to Euler angles.

Furthermore, quaternions are resistant to concerns such as singularities and discontinuities, which are possible problems when utilizing Euler angles. This durability is especially beneficial in cases where accurate and continuous attitude control is needed, such as in spaceship operations and aeronautical engineering. Moreover, quaternions are very effective in terms of processing resources, which is a key element in real-time systems where computational efficiency is vital.

In reality, the decision between Euler Angles and Quaternions relies on the individual application and the needs of the system. Euler Angles are generally selected for their ease of comprehension and visualization, particularly in educational environments or where human interaction is needed. Quaternions, on the other hand, are vital in computing jobs and real-time systems owing to their numerical stability and efficiency.

In conclusion, Attitude Representation is a crucial part of celestial mechanics, and both Euler Angles and Quaternions play essential roles in properly expressing an object's orientation in space. Euler Angles give an intuitive and visual depiction, but they are prone to gimbal lock. Quaternions, although more abstract, give numerical stability and efficiency, making them vital for computational applications and real-time control systems. The decision between the two strategies ultimately relies on the unique needs and limits of the application at hand.

8.3 Attitude Control Systems

Attitude control systems are the unsung heroes of space missions, responsible for the exact alignment and stability of spacecraft. In the harsh, weightless environment of space, keeping a desired orientation is vital for a myriad of activities including communication, solar power production, and scientific studies. These systems utilize a mix of hardware and complex algorithms to accomplish this feat.

One of the major components of an attitude control system is the collection of sensors that give information about the spacecraft's orientation relative to a reference frame. Inertial Measurement Units (IMUs) equipped with accelerometers and gyroscopes are the workhorses in this sector. Accelerometers detect linear acceleration, enabling the measurement of forces acting on the spacecraft. Gyroscopes, on the other hand, measure angular rates, allowing exact monitoring of rotational motion. These sensors operate in unison to deliver real-time data to the spacecraft's control algorithms.

Complementing the sensor suite are the actuators, which execute the instructions provided by the control algorithms. Reaction wheels and thrusters are the two basic kinds of actuators utilized in attitude control. Reaction wheels function on the concept of conservation of angular momentum. By rotating in the opposite direction to the required rotation, they generate a counteracting force on the spacecraft.

Thrusters, on the other hand, eject mass in one direction to create thrust in the opposing direction, so inducing controlled rotations. The choice between these actuators relies on mission needs, power limits, and spacecraft mass.

The control algorithms comprise the brain of the attitude control system, responsible for interpreting sensor data and creating instructions for the actuators. These algorithms involve a mix of classical control theory, state-space approaches, and sophisticated control techniques such as Kalman filtering. Proportional-Integral-Derivative (PID) controllers are often employed for their simplicity and efficacy in stabilizing a system. More complicated algorithms like the Extended Kalman Filter (EKF) are applied for accurate estimate of spacecraft attitude in the face of sensor noise.

Redundancy and fault tolerance are essential features of attitude control system design. Spacecraft commonly carry backup sensors and actuators to offset the impact of malfunctions. Moreover, the control software is often developed with fault detection and isolation (FDI) capabilities. In the case of a component failure, the FDI system may rearrange the control system to employ the remaining working hardware, assuring the mission's continuous success.

In conclusion, attitude control systems are the unseen hands that navigate spacecraft across the universe. Through the integration of sensors, actuators, and complex control algorithms, they guarantee the spacecraft retains its proper orientation, allowing it to fulfill its scientific goals. Redundancy and fault tolerance measures further boost its dependability. As we embark on increasingly ambitious space exploration projects, the refining and invention of attitude control systems will remain important in our attempt to traverse the immense expanse of the cosmos.

Chapter 9: Satellite Orbits and Space Debris

9.1 Low Earth Orbits (LEO), Medium Earth Orbits (MEO), Geostationary Orbits (GEO)

Low Earth Orbits (LEO) constitute a group of orbits placed very near to Earth's surface, generally ranging from heights of roughly 160 kilometers to 2,000 kilometers above sea level. One of the distinctive qualities of LEOs is their closeness, which provides various practical benefits. Satellites positioned in LEO benefit from decreased latency in communication and observation, making them excellent for applications including Earth observation, remote sensing, and satellite-based internet services. Moreover, because to their comparatively lower altitudes, LEOs display quicker orbital periods, allowing for more frequent revisits over specified geographical areas. This makes them important for monitoring dynamic phenomena such as weather patterns, natural catastrophes, and climate change. However, the negative of LEOs is their increased atmospheric drag, forcing periodic changes to maintain their orbital characteristics.

Medium Earth Orbits (MEO) occupy an area of space at an altitude ranging from roughly 2,000 kilometers to 35,786 kilometers above Earth's surface. The most well-known type of MEO orbits is the navigation satellite constellations like the Global Positioning System (GPS) and GLONASS. MEO satellites provide a balance between coverage area and orbital period. While they have somewhat greater latencies compared to LEOs, they maintain a consistent presence across a broader geographical region, making them vital for global navigation, communication, and geodetic applications. Additionally, MEO satellites encounter less air pull compared to LEOs, lowering the frequency of required orbital adjustments. The stability and dependability of MEO orbits have made them a cornerstone of contemporary telecommunications and navigation systems.

Geostationary Orbits (GEO) are a separate type of orbits positioned at an altitude of roughly 35,786 kilometers above the equator. These orbits possess a unique characteristic: satellites in GEO stay stable relative to a given point on the Earth's surface, providing for continuous surveillance of a specific area. This capability is of critical value for applications like weather monitoring, television broadcasting, and telecommunications. However, it's vital to remember that GEO satellites are more costly to launch and maintain owing to the large energy needed to reach this height. Additionally, the restricted amount of available slots in GEO might lead to congestion and coordination issues. Despite these restrictions, the unmatched reliability and coverage given by GEO satellites make them a vital component of the current space infrastructure.

In conclusion, each form of orbit - Low Earth Orbit (LEO), Medium Earth Orbit (MEO), and Geostationary Orbit (GEO) - has various benefits and is suitable to certain purposes. LEOs offer closeness and frequent revisits, making them excellent for Earth observation and remote sensing. MEOs find a compromise between coverage area and orbital period, making them vital for global navigation and communication systems. On the other hand, GEOs give continuous coverage across a specific territory, making them ideal for applications needing continual communication and monitoring. Understanding the features and uses of these orbits is crucial to the design and operation of satellites and space missions.

9.2 Molniya Orbits and Sun-Synchronous Orbits

Molniya orbits offer a separate type of very elliptical orbits that find wide applicability in satellite systems needing sustained coverage of high-latitude areas. Named after the Soviet Molniya communication satellites, these orbits are distinguished by their high eccentricity and inclination. Specifically, a Molniya orbit has an eccentricity close to 0.7, which means it deviates greatly from a complete circle, resulting in a huge gap between its apogee (highest point) and perigee (lowest point). This extended design means that the satellite spends the bulk of its time at high latitudes, where it achieves a slower seeming velocity across the sky.

The fundamental value of Molniya orbits is in their capacity to retain a relatively stable location over certain places of interest on the Earth's surface, especially those at higher latitudes. This is especially vital for applications such as communication, weather monitoring, and navigation, where continuous coverage of polar or near-polar areas is necessary. Molniya orbits are well-suited for such duties because to their distinctive shape, enabling satellites to loiter over these areas for lengthy durations.

Sun-synchronous orbits, also known as heliosynchronous or polar orbits, serve a fundamentally different function compared to Molniya orbits. These orbits are precisely engineered to coordinate with the location of the Sun in the sky, maintaining constant illumination conditions over the course of a year. Sun-synchronous orbits are defined by a precise inclination and altitude that allows the satellite to cross the equator at nearly the same local solar time on each orbit. This implies that the satellite travels over any particular spot on Earth at a regular time of day, regardless of the season.

This synchronous interaction with the Sun is vital for missions such as Earth observation and remote sensing. By guaranteeing continuous lighting conditions, satellites in sun-synchronous orbits can record pictures and data with uniform lighting and shadow conditions, making them particularly relevant for applications like environmental monitoring, agriculture, and climate research. Moreover, this synchronized timing enables for easier comparison of data acquired over time, aiding the discovery of long-term trends and changes in the environment.

In contrast to Molniya orbits, which are meant to concentrate on specific high-latitude areas, sun-synchronous orbits give a larger global view, making them especially suited for applications that need extensive and systematic Earth monitoring. The capacity to acquire data under stable illumination conditions throughout the year boosts the value of satellites in these orbits for scientific study and practical applications.

9.3 Space Debris Mitigation and Tracking

Space trash Mitigation and Tracking have become crucial parts of space operations due to the increased presence of human-made trash in Earth's orbit. This problem emerges from decades of space exploration, satellite launches, and in-orbit activity. The existence of space debris presents a substantial hazard to both functioning satellites and crewed missions in orbit. As such, it is necessary to assess the techniques and technology involved in space debris abatement and tracking.

Mitigating space debris entails minimizing the generation of new debris and lowering the danger of collisions with existing objects. One significant method is to build satellites with disposal plans, ensuring they are either transferred to a cemetery orbit at the end of their active life or deorbited safely into the Earth's atmosphere where they burn up. Additionally, the inclusion of propulsion systems on satellites may permit controlled re-entry or station-keeping operations, lowering the probability of unintentional accidents.

Furthermore, creating norms and standards for prudent space activities is vital. International organizations, such as the United Nations Office for Outer Space Affairs (UNOOSA), play a significant role in creating and promoting space debris reduction rules. These rules contain advice for spacecraft design, launch operations, and end-of-life disposal protocols. By adhering to these best practices, the space-faring community may collaboratively strive towards a safer orbital environment.

Tracking space debris is equally crucial for assuring the safety of space missions. Advanced monitoring systems, including as ground-based radars and telescopes, monitor objects in orbit, giving critical data on their trajectories, velocities, and possible collision dangers. Organizations like the U.S. Space Surveillance Network (SSN) and international consortia monitor thousands of active spacecraft, dead satellites, and chunks of debris, allowing space authorities and operators to make educated judgments about collision avoidance operations.

Moreover, the advancement of space situational awareness (SSA) technology is boosting our capacity to detect and anticipate the motions of space debris. SSA systems employ a mix of ground-based and space-based sensors to continually monitor the space environment. This information is then utilized to offer early warning of probable collisions, enabling operators to alter satellite orbits or take evasive action as required.

In recent years, there have been developments in autonomous collision avoidance systems, which allow satellites to make real-time judgments based on tracking data. These solutions have the ability to significantly minimize the danger of accidents and increase the overall safety of space operations.

In conclusion, space debris reduction and tracking are key components of responsible space operations.

Through careful satellite design, adherence to international norms, and the employment of modern monitoring and situational awareness technology, the space-faring community may work cooperatively to limit the threats presented by space debris. As the number of objects in orbit continues to expand, continuous study and development in this subject will be important for maintaining the sustainability and safety of space operations for future generations.

Chapter 10: Celestial Mechanics in Astrophysics

10.1 Binary Star Systems and Stellar Orbits

"Binary Star Systems and Stellar Orbits" is a significant chapter in celestial mechanics, offering insight on the delicate dance of stars in mutual gravitational embrace. This chapter digs into the remarkable phenomena of binary star systems, where two stars circle around a shared center of mass. Such systems offer a unique laboratory for studying celestial mechanics owing to their well-defined orbits and gravitational interactions.

In a binary star system, each star circles a shared barycenter, the point around which both stars spin. The nature of these orbits may vary significantly, from almost circular to severely elliptical. This variation in orbital properties leads to the complexity of binary star systems, allowing for a diverse array of observable events.

One of the fundamental elements impacting the dynamics of binary star systems is the mass ratio between the two stars. When the masses are comparable, the stars trace out observable orbits around the barycenter. However, in circumstances when one star greatly outweighs the other, the barycenter may be placed near to the more massive star, essentially turning the motion of the lighter star more analogous to a planet around a star. This instance shows the many configurations that binary star systems may assume.

Stellar orbits inside binary systems are tightly connected to Kepler's Laws of Planetary Motion. Observations of the periodic fluctuation in stellar brightness (known as light curves) allow astronomers to extract essential information about the system's geometry, such as the inclination angle of the orbit and the relative sizes of the stars. These observations, together with exact measurements of the stars' radial velocities, constitute the basis for understanding the dynamics of binary systems.

Moreover, binary star systems are crucial in astrophysics study for various reasons. They give a useful technique of calculating the masses of stars, a notoriously tough process in astrophysics. By monitoring the movements of stars in a binary system, astronomers may combine Newton's equations of motion and the law of universal gravitation to determine the masses involved. This insight, in turn, has far-reaching consequences for our understanding of star development and the larger dynamics of galaxies.

In conclusion, "Binary Star Systems and Stellar Orbits" is a vital chapter in celestial mechanics, giving a glimpse into the complexity of gravitationally coupled stellar pairs. By examining these systems, scientists gain insight into basic laws of celestial motion, enhancing our knowledge of star dynamics and the greater universe. Moreover, the consequences of this study extend beyond binary systems, revealing significant insights into the behavior of celestial objects in many astrophysical environments.

10.2 Galactic Dynamics and Dark Matter

"Galactic Dynamics and Dark Matter" dives into the complicated interaction between celestial entities inside galaxies and the mysterious nature of dark matter, a material that forms a major fraction of the universe but remains elusive to direct discovery. This area of research is fundamental in our knowledge of the universe, yielding vital insights into the origin, development, and behavior of galaxies over cosmic timescales.

Galactic dynamics, a basic component of this field, analyzes the motion and interactions of stars, gas, and other celestial entities inside galaxies. Newtonian physics and the law of gravity play a crucial role in these investigations, enabling scientists to describe and forecast the complicated orbits and interactions that regulate the behavior of galactic systems. Through rigorous observations and computer simulations, astronomers untangle the complex ballet of stars as they circle galactic centers, sometimes generating structures like as spiral arms, galactic bars, and even galactic clusters.

However, galactic dynamics alone does not account for all known events inside galaxies. One of the most baffling mysteries in current astrophysics is the existence of dark matter, a material that exerts gravitational effect on visible matter yet emits no detectable radiation. Its existence is inferred from a variety of astrophysical evidence, including galaxy rotation curves, gravitational lensing, and large-scale cosmic structures. This riddle has led to substantial theoretical and experimental attempts to explain its nature and spread.

The function of dark matter in galactic dynamics is significant. Observations of galaxy rotation curves, for instance, suggest that stars and gas near the edges of galaxies travel at very high velocities. This suggests the presence of unseen mass, scattered more broadly than the light matter can account for. Dark matter is hypothesized to create huge halos surrounding galaxies, giving the extra gravitational force required to explain these observed velocities.

Moreover, dark matter's existence is important in the birth and development of galaxies. Through gravitational interactions, dark matter halos provide the framework upon which visible matter condenses, permitting the development of galaxies as we know them. Understanding the distribution and characteristics of dark matter is therefore important to grasping the larger-scale structures and dynamics of our cosmos.

In essence, "Galactic Dynamics and Dark Matter" is a multidimensional area of research that blends traditional celestial mechanics with a desire to fathom the mysterious, non-luminous material that pervades the universe. By merging data, computer models, and theoretical frameworks, scientists attempt to understand the complex dance of celestial entities inside galaxies and discover the basic role dark matter plays in sculpting the universe at galactic and cosmic scales. This initiative stands as a cornerstone in contemporary astrophysics, promising to expand our knowledge of the universe's history, present, and future.

10.3 Celestial Mechanics in Exoplanetary Systems

The study of celestial mechanics within exoplanetary systems is an emerging area that tries to explain the complicated dynamics regulating planets orbiting stars outside our solar system. With the discovery of hundreds of exoplanets in recent decades, knowing the subtleties of their orbital behavior has become important in describing these faraway worlds. Celestial mechanics, a cornerstone of astrodynamics, offers the theoretical foundation for grasping the motion of celestial bodies. When applied to exoplanetary systems, it gives insight on many phenomena including planetary orbits, resonances, and stability areas.

One of the major components of celestial mechanics in exoplanetary systems focuses around the calculation of orbital parameters. Unlike our familiar solar system, exoplanetary systems display a remarkable number of orbital patterns, from close-in hot Jupiters to distant ice exoplanets. This variety challenges existing celestial mechanics models and necessitates updated theories to account for the unusual and frequently unexpected planetary orbits. Understanding semi-major axes, eccentricities, and inclinations in exoplanetary systems not only assists in defining individual planets but also gives crucial insights into the creation and evolutionary processes of these systems.

Resonances, a crucial concept in celestial physics, play a vital role in exoplanetary systems. These resonant interactions occur when two or more celestial bodies exert periodic effects on one other, resulting to stable orbital configurations. Detecting resonances in exoplanetary systems gives vital information about their dynamical histories and probable migration pathways. For instance, the existence of mean-motion resonances might suggest to early planetary migration, revealing insights about the system's creation and subsequent development.

Stability areas constitute another key feature of celestial mechanics in exoplanetary systems. These areas outline the parameter space where planets may sustain stable orbits over lengthy periods. Analyzing stability areas helps find livable zones where circumstances may be favorable to the existence of life. Moreover, it informs our knowledge of the long-term stability of exoplanetary systems, which is vital for appraising their potential for harboring life or for future human exploration.

Challenges in the study of celestial mechanics in exoplanetary systems continue. The variety of observed exoplanetary systems needs the development of specific models and computational tools. Furthermore, the complexity imposed by multi-planet systems and the possible effect of stellar partners demands extensive numerical models to adequately forecast their long-term behavior.

In conclusion, celestial mechanics in exoplanetary systems sits at the forefront of current astrodynamics.

By applying the concepts of celestial mechanics to the many exoplanetary systems found, scientists attempt to comprehend the complexity of planetary orbits, resonances, and stability areas. This undertaking not only increases our knowledge of the dynamics of distant planets but also bears substantial implications for the hunt for livable settings and the possibility for alien life. As the science continues to expand, breakthroughs in computational tools and strengthened theoretical frameworks promise to open more insights into the celestial mechanics of exoplanetary systems.

Chapter 11: Celestial Mechanics in Space Missions

11.1 Trajectory Design for Interplanetary Missions

"Trajectory Design for Interplanetary Missions" comprises a fundamental element of celestial mechanics, concentrating on the intricacy of charting routes for spacecraft to travel between planets in our solar system. This discipline includes the use of astrodynamics, gravitational assistance, and mathematical modeling to discover the most efficient and effective trajectories. One of the primary issues in interplanetary mission design comes in minimizing energy consumption while fulfilling the required mission goals.

A crucial factor in trajectory design is the employment of gravity aids. By precisely scheduling a spacecraft's contact with a planet or moon, engineers may harness the body's gravitational pull to deliver an extra boost in velocity. This technology allows for more efficient trajectories, decreasing the quantity of fuel needed and allowing missions to reach their targets with faster speed. Gravity assistance have been applied in several successful missions, notably the Voyager and Cassini missions, which leveraged flybys of multiple planets to reach their final destinations in the outer solar system.

Moreover, trajectory planning for interplanetary missions demands a full grasp of celestial physics and the complicated interactions between celestial entities. This requires the computation of orbital components, identification of launch windows, and consideration of planetary alignments. Engineers must account in the varied locations and velocities of planets along their separate orbits, attempting to uncover chances for ideal mission routes. Such calculations demand extensive numerical techniques and simulations to effectively represent the behavior of spacecraft in the dynamic environment of the solar system.

Another crucial part of trajectory design is mission duration. Engineers must achieve a compromise between the demand for quick travels and the restrictions imposed by current propulsion systems. Longer missions may need smaller propulsion systems, decreasing launch costs and complexity, but may confront issues in sustaining spacecraft operation over long periods. Conversely, shorter flights may need strong propulsion systems, thereby restricting spacecraft size and complicating mission logistics. Striking this equilibrium is vital for assuring mission success.

Furthermore, trajectory planning for interplanetary missions is directly connected to the idea of stability and robustness. Engineers must consider possible causes of disturbance, such as solar radiation pressure, gravitational anomalies, and even interactions with minor celestial bodies like asteroids. These variables may create variations from the predicted trajectory, requiring the performance of corrective maneuvers or changes throughout the flight. Anticipating and managing such disruptions is vital for securing the success of interplanetary missions.

In conclusion, "Trajectory Design for Interplanetary Missions" is a multidisciplinary topic that uses concepts of astrodynamics, celestial mechanics, and mission planning to map the most efficient and effective pathways for spacecraft traveling between planets. The employment of gravity aids, exact mathematical modeling, and consideration of mission length and stability are important to this undertaking. Engineers in this sector play a crucial role in allowing successful missions to explore and study the different ecosystems of our solar system. Their effort not only increases our scientific knowledge but also prepares the path for future initiatives in space travel.

11.2 Gravity Assist Maneuvers

Gravity assist movements, commonly known as gravitational slingshots or flybys, constitute an essential and clever tool in celestial mechanics and space exploration. This maneuver takes use of the gravitational attraction of celestial entities, often planets or moons, to modify the trajectory and velocity of a spacecraft. It has been deployed in multiple space missions, changing the efficiency of interplanetary transport and deep space exploration.

The main premise underlying gravity assist movements is the conservation of energy and angular momentum. As a spacecraft approaches a big celestial body, such as a planet, it enters the body's gravitational field. During this phase, the spacecraft's velocity relative to the planet rises, considerably raising its kinetic energy. This increase in speed occurs at the price of the planet's own orbital energy, keeping to the rule of conservation of angular momentum.

Consequently, the spacecraft escapes the encounter with increased velocity, enabling it to reach distant destinations with less energy usage.

One of the most notable instances of gravity assist techniques is NASA's Voyager mission. Voyager 1 and 2 exploited gravity assistance to investigate the farthest planets of the solar system. By exploiting Jupiter's gravity in the 1970s and Saturn's pull in the early 1980s, these spacecraft greatly improved their velocity, allowing them to reach Uranus and Neptune in less time and with reduced fuel needs. Such missions wouldn't have been achievable without these tactics.

The key to the effectiveness of gravity assist operations is meticulous preparation. Mission planners must carefully pick the celestial bodies to be utilized for the assist, considering their locations in space and the spacecraft's trajectory. Moreover, the time of the maneuver is essential, since it demands synchronizing the spacecraft's course with the target planet's location. This degree of accuracy underscores the multidisciplinary nature of celestial mechanics, necessitating complicated calculations and interaction between celestial navigation specialists and mission planners.

Gravity assist maneuvers have not only shortened mission length and expense but have also allowed for several scientific investigations during the flyby of a celestial body. For example, the Cassini-Huygens mission to Saturn performed multiple close flybys of Saturn's moon Titan, acquiring crucial data and photos before finally entering Saturn's orbit. This multi-purpose approach to exploration has been a characteristic of gravity assistance.

In conclusion, gravity assist maneuvers serve as a wonderful tribute to human inventiveness in space exploration. By harnessing the gravitational pulls of celestial planets, these maneuvers have expanded the reach of our space probes and have led to astounding discoveries across the solar system and beyond. They continue to play an important role in the realm of celestial mechanics and space exploration, giving effective means for reaching distant locations and performing scientific study in the universe.

11.3 Orbital Insertion and Rendezvous

"Orbital Insertion and Rendezvous" are crucial maneuvers in space missions, allowing spacecraft to acquire exact orbits and rendezvous with other celestial entities. Orbital insertion includes the act of inserting a spacecraft into its target orbit around a celestial body, whereas rendezvous encompasses the controlled approach and meeting of two or more spacecraft in space. These moves involve cautious preparation, perfect execution, and a sophisticated grasp of celestial physics.

Orbital insertion is a critical step of every space mission, setting the route and altitude at which a spacecraft will operate. This maneuver requires a mix of engine burns and modifications to the spacecraft's velocity and orientation. The aim is to establish a stable orbit that satisfies the mission goals, whether it be for Earth observation, planetary exploration, or satellite placement. Engineers precisely calculate the required force and timing of burns to ensure the spacecraft reaches its intended orbit with maximum accuracy. Any variation from the predicted trajectory might have major repercussions, underlining the need of precision orbital insertion.

Rendezvous, on the other hand, is a difficult procedure requiring a great degree of accuracy and cooperation. It is deployed in cases when two or more spacecraft need to come together in orbit, such as in personnel transfer, satellite maintenance, or research missions involving numerous spacecraft. Achieving rendezvous demands a meticulous choreography of orbital movements, taking into consideration the relative locations, velocities, and orbital components of the spacecraft involved. Furthermore, the regulation of relative speeds and the alignment of spacecraft axes are critical to achieve a safe and precise rendezvous.

The mathematical underpinnings of celestial mechanics play a key role in both orbital insertion and rendezvous procedures. Understanding the orbital components and disturbances that impact the spacecraft's trajectory is vital in computing the exact maneuvers necessary. Perturbations from gravitational pulls of astronomical bodies, atmospheric drag, and other variables must be accounted for to guarantee the spacecraft reaches its intended destination precisely.

Moreover, developments in astrodynamics and computational tools have considerably boosted the precision and efficiency of orbital insertion and rendezvous procedures. Advanced algorithms and numerical approaches allow mission planners to model and optimize these movements, taking into account different mission limitations and uncertainties. This enables for more exact orbital insertions and rendezvous, even in extremely complicated and dynamic situations.

In conclusion, orbital insertion and rendezvous are crucial components of space missions, requiring a comprehensive grasp of celestial physics and rigorous preparation. The proper execution of these maneuvers is vital for attaining mission goals, whether it includes putting a satellite into its allotted orbit or coordinating a rendezvous between many spacecraft. With the constant progress of astrodynamics and computational tools, the accuracy and efficiency of these maneuvers continue to increase, opening the path for increasingly ambitious and complicated space exploration activities.

Chapter 12: Advanced Topics: Chaos and Resonance

12.1 Chaos in Celestial Mechanics

Chaos in celestial mechanics is a phenomena that occurs from the complex, non-linear interactions between celestial bodies inside a gravitational system. This hypothesis, founded in the work of Henri Poincaré in the late 19th century, transformed our knowledge of celestial dynamics. Prior to this discovery, it was popularly assumed that the motion of celestial bodies followed deterministic, predictable routes. However, Poincaré's investigations indicated that even in apparently simple three-body systems, minute alterations in beginning circumstances might lead to dramatically divergent trajectories over time.

One of the major properties of chaotic systems in celestial mechanics is sensitivity to beginning circumstances, frequently referred to as the "butterfly effect." This implies that even slight changes in the initial locations or velocities of celestial bodies may lead to significantly different results in their long-term behavior. For example, in a three-body system, if two planets start in relatively close proximity with virtually similar beginning circumstances, their future orbits might diverge dramatically over time, leading to radically different configurations.

Chaos in celestial mechanics has major ramifications for our capacity to make long-term predictions regarding the behavior of celestial bodies. It establishes fundamental restrictions on the accuracy with which we can anticipate the locations and velocities of planets, asteroids, and comets over lengthy periods. This has practical repercussions for space missions, where accurate trajectory planning and guidance are vital. Mission designers must carefully evaluate the uncertainty generated by chaotic behavior when mapping paths for spacecraft.

Moreover, chaotic behavior in celestial mechanics has led to the discovery of intriguing dynamical structures, including as chaotic zones and stable resonances. These areas in phase space may hold a mix of stable and unstable orbits, contributing to the diverse tapestry of movements inside a gravitational system. For instance, the Kozai-Lidov mechanism is a well-known example of a chaotic resonance that influences the orbits of asteroids and satellites, creating oscillations in eccentricity and inclination.

Chaos also has ramifications for our greater knowledge of the universe. It contradicts the reductionist concept that complex systems can always be studied and predicted based on beginning circumstances alone. Instead, it argues that certain systems may be fundamentally unpredictable over extended time spans. This revelation has generated interest in the study of non-linear dynamics and chaos theory across different scientific areas, going well beyond celestial mechanics.

In conclusion, chaos in celestial mechanics constitutes a major break from past deterministic theories of celestial motion. It provides a degree of unpredictability and sensitivity to beginning circumstances that has major practical and theoretical ramifications. Embracing chaos has enhanced our knowledge of the intricacies of gravitational systems and has opened up new paths of investigation in celestial dynamics and beyond. It serves as a tribute to the complicated and sometimes surprising nature of the universe.

12.2 Kozai-Lidov Mechanism

The Kozai-Lidov Mechanism, named after its discoverers Yoshihide Kozai and Michael Lidov, is a significant phenomena in celestial mechanics that explains the development of the orbital components of a tiny body in a hierarchical, three-body system. This process often includes a tiny, inclined body circling a bigger core body in the presence of a third, much more massive body, such as a planet or star.

The essential characteristic of the Kozai-Lidov Mechanism is its capacity to produce considerable fluctuations in both the eccentricity and inclination of the tiny body's orbit over lengthy periods of time. These fluctuations occur regularly, causing the orbit to shift between states of high eccentricity and low inclination, and vice versa. This behavior is especially significant in the study of many celestial objects, including asteroids, comets, and even exoplanetary systems.

One of the distinguishing aspects of the Kozai-Lidov Mechanism is the conservation of a quantity known as the Kozai constant. This constant indicates the connection between the eccentricity and inclination of the tiny body's orbit. As the orbit develops under the influence of the third, huge body, the Kozai constant stays constant. This conservation principle allows for the prediction and study of the orbital development of the tiny body over extended timeframes.

Furthermore, the Kozai-Lidov Mechanism has substantial consequences for the behavior of celestial bodies inside our solar system. For instance, it plays a vital role in the development of objects in the outer solar system, such as the trans-Neptunian objects and certain kinds of asteroids. Additionally, the Kozai-Lidov Mechanism has been seen in binary star systems, where it may lead to significant changes in the eccentricity and inclination of the orbit of one star around the other.

Despite its theoretical relevance, it's vital to note that the Kozai-Lidov Mechanism is impacted by several things, such as perturbations from extra bodies, relativistic effects, and higher-order gravitational interactions. These intricacies may make exact predictions problematic, requiring the use of sophisticated numerical tools and simulations for reliable modeling.

In summary, the Kozai-Lidov Mechanism is a key idea in celestial mechanics that defines the subtle interaction between eccentricity and inclination in a three-body system. Its applications vary from studying the dynamics of asteroids and comets to giving insight on the behavior of binary star systems. While theoretical in nature, its real-world consequences are substantial, giving vital insights into the long-term development of celestial objects in our universe.

12.3 Arnold Diffusion and Stability

Arnold Diffusion is a phenomena in celestial mechanics that plays a vital role in understanding the long-term behavior of dynamic systems with numerous degrees of freedom, especially in the context of chaotic orbits. Named for the great mathematician Vladimir Arnold, this theory gives insights into the stability and predictability of celestial entities under a changing gravitational environment.

At its foundation, Arnold Diffusion centers around the idea of disturbances in astronomical orbits. In a multi-body system, such as planets circling a star or satellites orbiting a planet, interactions between celestial bodies generate perturbations in the orbits. These disturbances may lead to major modifications in the trajectories over time, resulting in what is known as chaotic behavior. Arnold Diffusion helps us understand the circumstances under which such chaotic behavior might lead to orbital instability.

One of the important contributions of Arnold Diffusion is the explication of the KAM theorem, named after Kolmogorov, Arnold, and Moser. This theorem illustrates that under certain circumstances, perturbations to regular orbits, known as KAM tori, may lead to chaotic motion. This phenomena is especially significant in celestial mechanics because even slight perturbations, such as those from other celestial bodies or non-spherical forms, may have enormous impacts over lengthy periods.

In practical terms, Arnold Diffusion has substantial ramifications for space missions and satellite operations. It underlines the need of precise and accurate orbit determination and control. Understanding the possibility for chaotic behavior helps mission planners to predict and limit the consequences of disturbances, assuring the long-term stability of spacecraft.

Moreover, Arnold Diffusion has larger implications outside celestial mechanics. It finds importance in many domains like as plasma physics, molecular dynamics, and even in economic and social sciences where complex systems with numerous interacting constituents are addressed.

While Arnold Diffusion gives useful insights into the behavior of dynamic systems, it also creates issues in terms of predictability. The fundamentally chaotic structure of some orbits suggests that long-term forecasting may be restricted, demanding regular monitoring and modifications. This has practical ramifications for the planning and operation of space missions, particularly those requiring long-duration voyages or interactions with several celestial bodies.

In conclusion, Arnold Diffusion stands as a basic idea in celestial mechanics, offering insight on the complicated dance of celestial entities under dynamic gravitational environments. Its explication of chaotic behavior and its influence on long-term stability has far-reaching consequences for space exploration and satellite operations. While it creates issues in predictability, it also underlines the significance of accuracy in orbit determination and control. As our knowledge of celestial physics continues to grow, Arnold Diffusion remains a cornerstone in comprehending the intricacies of our cosmic environment.

Chapter 13: Celestial Navigation and Space Exploration

13.1 Astrodynamics in Space Navigation

Astrodynamics plays a vital role in contemporary space navigation, offering the theoretical foundation and practical procedures for mapping exact paths across the universe. This subject integrates ideas of celestial physics and engineering to correctly forecast and control the velocity of spacecraft. By understanding the complexity of gravitational interactions, orbital dynamics, and propulsion systems, astrodynamics facilitates the design and execution of complicated space missions.

At its foundation, astrodynamics depends on the basic rules of physics, notably Newton's equations of motion and the universal law of gravity. These principles constitute the foundation for computing the trajectories of spacecraft, considering elements like as beginning velocity, gravitational forces exerted by celestial bodies, and any extra push delivered by onboard propulsion systems. Through rigorous mathematical modeling, astrodynamists may simulate and forecast the route a spacecraft will take in space, allowing for exact mission planning.

One of the key parts of astrodynamics is the determination of orbital parameters. These factors, such as semi-major axis, eccentricity, and inclination, dictate the shape and direction of an orbit. Astrodynamists employ these factors to calculate and anticipate the future locations of spacecraft, ensuring they reach their intended destinations with accuracy. Additionally, astrodynamics assists in planning orbital transfers and rendezvous operations, maximizing the usage of propellants and decreasing trip time.

Astrodynamics is very crucial for interplanetary missions. These missions need rigorous planning to take advantage of celestial mechanics phenomena such as gravity aids, which entail exploiting the gravitational pull of planets to modify a spacecraft's trajectory and obtain additional velocity. This strategy is crucial in saving fuel and enhancing the efficiency of missions, making it a cornerstone of contemporary space exploration.

Moreover, astrodynamics plays a key role in the precision of communication with spacecraft. Deep space missions depend on elaborate networks of ground-based antennas, known as the Deep Space Network (DSN), to maintain communication with spacecraft. Astrodynamics helps anticipate the locations of both the spacecraft and Earth relative to one other, allowing for exact scheduling of communication windows. This guarantees that data transfer happens with minimum delays and disturbances.

In recent years, the science of astrodynamics has undergone considerable breakthroughs driven by technical innovation and computing capacity. High-fidelity models and improved numerical approaches now allow for more precise forecasts of spacecraft trajectories, allowing planners to account for complicated gravitational interactions and orbital disturbances. Additionally, the integration of artificial intelligence and machine learning methods is revolutionizing autonomous navigation systems, giving spacecraft with the power to make real-time alterations to their trajectories depending on observed circumstances.

In conclusion, astrodynamics serves as the backbone of space navigation, merging concepts of celestial mechanics and engineering to allow exact planning and execution of space missions. Through the use of Newton's principles and the knowledge of orbital dynamics, astrodynamists may precisely anticipate the motion of spacecraft, create ideal routes, and harness celestial mechanics phenomena for efficient interplanetary travel. As technology continues to progress, the area of astrodynamics is set to play an even bigger role in influencing the future of space travel.

13.2 Deep Space Network and Communication Delays

The Deep Space Network (DSN) serves as a cornerstone of interplanetary exploration, providing crucial communication between Earth and distant spacecraft. Comprising a network of ground-based antennas strategically positioned throughout the world, the DSN permits a continuous contact with spacecraft reaching far beyond Earth's orbit. However, one key factor in this sophisticated communication process is the problem of communication delays.

Communication delays within the context of the DSN relate to the lag in signal transmission between Earth and spacecraft operating in deep space. This delay is mostly owing to the great distances that separate these heavenly objects. Due to the limiting speed of light, signals take a great amount of time to travel these huge distances. Consequently, this causes a delay that may vary from several minutes to hours, depending on the spacecraft's distance from Earth.

The repercussions of these communication delays are enormous, especially in mission planning and execution. Real-time control and decision-making become unfeasible, needing a highly autonomous spaceship capable of performing pre-programmed instructions in reaction to unanticipated occurrences. This autonomy is vital for missions to other planets, since standard remote control from Earth is impracticable because to the lengthy connection latency.

Moreover, the communication delays provide a barrier for missions requiring precise coordination, such as orbital maneuvers or landing procedures. Engineers and scientists must precisely prepare these movements, factoring in the delay to guarantee that orders are received and implemented properly. This accuracy is crucial, since any mistake might have far-reaching ramifications for the mission's success.

Despite the issues caused by transmission delays, the DSN incorporates modern technology and approaches to limit their effect. Forward error correcting methods and complex modulation systems are employed to maintain the integrity of data transferred across large distances. Additionally, the DSN features a phased-array antenna technology, permitting simultaneous communication with numerous spacecraft, thereby enhancing its operating efficiency.

Furthermore, the DSN's strategic placing of ground stations in different regions across the globe assures continuous coverage regardless of the spacecraft's position in its orbit. This worldwide network successfully avoids blackout times and optimizes the quantity of data that can be delivered and received within a given communication session.

In conclusion, the Deep Space Network and the related communication delays are crucial components of interplanetary exploration. The constraints created by these delays entail a high degree of autonomy in spacecraft operations and demand thorough preparation for important maneuvers. Through modern technology and a worldwide network of ground stations, the DSN effectively navigates these hurdles, allowing mankind to connect with and explore the outer reaches of our solar system and beyond.

13.3 Future of Space Exploration: Missions to Mars, Asteroids, and Beyond

The future of space exploration is an exciting frontier that comprises trips to places such as Mars, asteroids, and beyond. This trajectory indicates a dramatic change in humanity's pursuit for knowledge and comprehension of the universe. Mars, frequently referred to as the "Red Planet," has caught the curiosity of scientists and science fans alike. With breakthroughs in propulsion technology and spaceship design, trips to Mars are becoming more possible. The ultimate objective is to establish a human presence on Mars, which involves a range of obstacles including long-duration space flight, life support systems, and radiation shielding. The success of such missions rests on our capacity to design sustainable dwellings and technology that can resist the hostile Martian environment.

Asteroids, the leftovers of our early solar system, are another fascinating object for investigation. These celestial bodies carry vital information about the beginnings of our solar system and the potential for precious resources like metals and water. Missions to asteroids include complicated rendezvous and sample return procedures, needing precision navigation and autonomous spacecraft capabilities. Additionally, knowing the composition and structure of asteroids is vital for planetary defense, since some of them have the potential to impact with Earth.

Looking beyond our near cosmic neighbors, interstellar travel is a focus of theoretical and technical investigation. While the hurdles are great, including the massive distances and the need for novel propulsion technologies, the quest of interstellar travel symbolizes the apex of human curiosity and ambition. Concepts like solar sails, nuclear propulsion, and advances in basic physics may hold the key to making interplanetary expeditions a reality. These initiatives would not only push the frontiers of our scientific knowledge but might possibly lead to the finding of alien life or totally new domains of astrophysical phenomena.

The future of space exploration also involves ethical and legal problems. As we journey deeper into space, considerations concerning resource use, environmental effect, and possible contacts with alien lifeforms become crucial. International collaboration and governance in space operations will play a significant role in ensuring the responsible and sustainable growth of humanity's presence beyond Earth.

In conclusion, the future of space exploration has great promise and possibility for humans. Missions to Mars, asteroids, and the possibility of interstellar travel represent the next frontier in our quest to comprehend the cosmos. These activities not only stretch the bounds of our technical capabilities but also need serious consideration of ethical, legal, and environmental repercussions. As we begin on this voyage, we stand at the brink of a new era in human exploration, one that has the potential to alter our knowledge of the universe and our role within it.

Chapter 14: Future Frontiers in Celestial Mechanics

14.1 Interstellar Travel and Orbits in Other Star Systems

"Interstellar Travel and Orbits in Other Star Systems" dives into the theoretical and practical obstacles of traveling beyond our solar system, an activity that marks the zenith of human space exploration objectives. This chapter navigates through the complicated terrain of astrodynamics, presenting tactics and conceptual frameworks for interstellar travel.

At its heart, interstellar travel demands a grasp of the astrodynamics of star systems beyond our own. The book covers the basic subject of how to shift from the familiar celestial mechanics of our solar system to the very different dynamics observed in other star systems. It grapples with topics such as the computation of escape velocity from stars, planetary motion around alien suns, and the idiosyncrasies of interstellar paths.

Central to the topic is the idea of "gravitational aids" or "slingshots", which use the gravitational fields of celestial bodies to modify the course of a spacecraft. This approach, notably deployed in missions like Voyager and New Horizons, is crucial for increasing efficiency in interstellar travel. The chapter includes thorough studies of how slingshot movements could be deployed in the context of star systems beyond our own, offering insights into the particular obstacles and possibilities given by each.

Moreover, the chapter interacts with potential technology that may permit interstellar travel. Concepts such as "light sails" and "nuclear pulse propulsion" being researched for their potential to attain the tremendous speeds necessary to reach other star systems within a human lifetime. The book presents extensive appraisals of the theoretical viability and practical applicability of these technologies, while noting the significant technical challenges that must be surmounted.

The chapter also discusses the long-duration aspect of interplanetary voyages. It analyzes the issues given by life support systems, crew psychology, and the sheer solitude of space travel that continues over decades or perhaps centuries. Discussions on sleep, closed-loop life support, and sophisticated propulsion systems are integrated throughout the story, allowing a full exploration of the biological and technical components of interstellar voyages.

Furthermore, the book examines the subtleties of orbital mechanics surrounding stars that may vary greatly from our Sun. It dives into the intricacies of orbital stability, gravitational interactions between celestial entities, and the prospective finding of exoplanets with habitats favorable to human life. By studying the different characteristics that determine orbits in these alien systems, the chapter sets the framework for comprehending the challenges of navigation and maneuvering in interstellar space.

In summary, "Interstellar Travel and Orbits in Other Star Systems" presents a complete analysis of the theoretical and practical concerns surrounding interstellar travel. By examining themes ranging from astrodynamics to sophisticated propulsion technologies and life support systems, the chapter presents a complete picture of the enormous difficulties and tantalizing potential of exploring beyond our solar system. It lays the groundwork for a future when interplanetary travel, long the domain of science fiction, may become a reality.

14.2 Lagrange Clouds and Advanced Celestial Navigation Techniques

"Lagrange Clouds and Advanced Celestial Navigation Techniques" digs into the intriguing area of celestial mechanics, investigating the complicated dynamics of Lagrange points and its applications in space navigation. This chapter presents an advanced perspective on the utilization of Lagrange points, which are unique positions in a gravitational system where the gravitational forces of two large bodies balance the centripetal force felt by a smaller object, allowing it to maintain a stable position relative to those bodies.

The Lagrange Clouds, a concept presented in this chapter, are areas around the Lagrange points, defined by their dynamic stability. These areas are of tremendous importance for space missions owing to their potential for long-term stable orbits, making them suitable places for spacecraft, space telescopes, or observatories. The book investigates the practical consequences of deploying Lagrange Clouds for diverse objectives, such as Earth monitoring, space exploration, and astronomical study.

Moreover, the chapter goes into the deep mathematics and calculations required in forecasting and navigating inside these Lagrange Clouds. Advanced celestial navigation systems are required for correctly guiding spacecraft inside these dynamically complicated settings. This entails utilizing high-fidelity models of the Earth-Moon-Sun system and applying advanced computational techniques to compute the trajectories and orbital dynamics of spacecraft operating near Lagrange points.

The chapter also examines the obstacles and implications connected with Lagrange Cloud navigation. Factors including as perturbations from other celestial bodies, station-keeping maneuvers, and spacecraft restrictions are fully studied. Additionally, it dives into the technical breakthroughs necessary to allow spacecraft to function efficiently inside the Lagrange Clouds, including sophisticated propulsion systems, autonomous navigation algorithms, and precise control systems.

Furthermore, the chapter covers real-world applications and missions that have used Lagrange points and their associated Lagrange Clouds. Notable examples would include the James Webb Space Telescope, which is designed to orbit the second Lagrange point (L2), or missions that have exploited Lagrange points for interplanetary exploration, such as the SOHO and LISA Pathfinder missions.

In conclusion, "Lagrange Clouds and Advanced Celestial Navigation Techniques" gives a complete analysis of the theoretical foundations and practical uses of Lagrange points and their related areas. By giving insights into the mathematical models, navigational tactics, and real-world missions, this chapter offers readers with a comprehensive grasp of how Lagrange Clouds play a key role in the navigation and exploration of space. This sophisticated viewpoint is useful for academics, engineers, and space enthusiasts looking to push the frontiers of celestial mechanics and space navigation.

14.3 Quantum Effects in Celestial Mechanics

"Quantum Effects in Celestial Mechanics" dives into the interesting junction of quantum physics and classical celestial mechanics, exposing the subtle but significant influence that quantum events have on the behavior of celestial bodies inside our cosmos. This topic, at first view, may appear like an unusual merging of disciplines, but it underlines the complicated structure of the universe and the necessity for a comprehensive approach to comprehending its dynamics.

One of the fundamental quantum phenomena investigated in this context is the uncertainty principle. As described by Werner Heisenberg, this basic principle says that some pairings of physical attributes, such as location and momentum, cannot be simultaneously accurately specified. Applied to celestial mechanics, this indicates that there is inherent uncertainty in the exact location and momentum of celestial bodies. While in classical physics, locations and momenta are viewed as certain, this quantum uncertainty provides a measure of unpredictability, particularly at the quantum scales relevant to celestial bodies.

Furthermore, the idea of wave-particle duality becomes a significant concern. Particles, especially heavenly bodies, may display both particle-like and wave-like behavior. This duality has ramifications for the behavior of celestial bodies, especially at the quantum level. Understanding the wave-like character of celestial bodies becomes vital for understanding phenomena like diffraction patterns in the interactions of photons with celestial objects, providing an extra layer of complexity to the usual study of celestial mechanics.

Moreover, the entanglement phenomena in quantum physics bears significant implications for celestial mechanics. Entanglement refers to the connectivity of particles, where the condition of one particle immediately impacts the state of another, regardless of the distance between them. In the context of celestial bodies, this shows that apparently isolated entities may have subtle, interrelated affects on one another, defying traditional views of gravitational interactions as completely local events.

Additionally, the quantization of energy levels in astronomical systems must be addressed. Classical celestial mechanics frequently regards the energy of a system as continuous, whereas at the quantum level, energy levels are quantized. This means that celestial systems may only hold specific discrete energy levels, leading to phenomena like quantized orbits and energy transitions, which have ramifications for the stability and behavior of celestial bodies.

In summary, "Quantum Effects in Celestial Mechanics" presents a comprehensive examination of the complicated relationship between quantum phenomena and conventional celestial mechanics. By introducing ideas such as the uncertainty principle, wave-particle duality, entanglement, and energy quantization, this discipline deepens our knowledge of celestial dynamics, shining light on hitherto undiscovered frontiers in the world of astrodynamics. Embracing the coexistence of quantum and classical physics is crucial for a thorough knowledge of the complexity guiding the behavior of celestial entities in our vast universe.

Chapter 15: Ethical Considerations and Space Law

15.1 Space Traffic Management and Collision Avoidance

Space traffic management and collision avoidance are key components of contemporary space activities. With a rising number of satellites, spacecraft, and space debris in Earth's orbit, efficient coordination and regulation are vital to maintain the safety and sustainability of space operations.

One of the primary issues in space traffic management is the sheer amount of objects in orbit. Thousands of functioning satellites, dead spacecraft, and chunks of junk from earlier missions inhabit different orbital regimes. This congested environment demands rigorous planning and cooperation to minimize accidents and assure the continuing performance of operating equipment.

To solve this problem, space organizations, regulatory authorities, and commercial companies are creating sophisticated tracking and monitoring technologies. Ground-based radars and space-based sensors detect objects in near-Earth orbit, delivering real-time positional data. This information is then utilized to forecast probable conjunctions or near approaches between objects, allowing for timely moves if required.

Additionally, space agencies throughout the globe are working on standardized communication methods and data-sharing agreements. These projects attempt to facilitate the transmission of information between various space operators and regulatory authorities, supporting a collaborative approach to space traffic management. Such collaboration is vital in averting misunderstandings and disputes, particularly in circumstances of close encounters.

Furthermore, collision avoidance systems involve a mix of orbital modifications and spacecraft movements. For instance, propulsion systems may be employed to modify an object's trajectory, either by increasing or lowering its orbit or altering its inclination. Alternatively, passive approaches like spinning a spacecraft to vary its cross-sectional area may also be deployed to lessen the risk of a collision.

In the event that a collision risk is considered too high to minimize using maneuvers alone, contingency planning and collision probability assessments play a key role. These plans may include advice for satellite operators to conduct evasive maneuvers or, in extreme instances, to commence controlled satellite deorbiting processes to securely dispose of deceased spacecraft or debris.

Additionally, regulatory agencies are exploring the installation of tougher criteria for satellite design and end-of-life disposal. These initiatives attempt to reduce the formation of new space trash and promote ethical space operations. They may include requirements for satellites to have propulsion systems for controlled descent or to relocate to a cemetery orbit at the end of their operational life.

In conclusion, space traffic management and collision avoidance are key features of current space operations. The sheer number of objects in orbit necessitates advanced monitoring, communication, and maneuvering capabilities. Through coordinated efforts, new technology, and regulatory measures, the space community aims to assure the safety and sustainability of operations in Earth's orbit, therefore preserving the future of space exploration and usage.

15.2 Space Debris Regulations

Space debris laws refer to the collection of norms and legal frameworks developed by international organizations and spacefaring governments to control and reduce the rising problem of space debris. As human operations in space have expanded over the years, so has the quantity of defunct and non-functional objects circling Earth. These vary from spent rocket stages to decommissioned satellites, presenting major threats to active spacecraft and possibly rising into a severe environmental issue. Consequently, space debris laws are vital in sustaining the sustainability of space operations and assuring the safety of both crewed and uncrewed missions.

One essential part of space debris laws includes the monitoring and tracking of objects in orbit. Various space agencies and organizations, like as NASA and the European Space Agency (ESA), run monitoring capabilities to monitor the motions of space debris and operational satellites. These facilities give critical information for collision avoidance operations and allow space operators to alter the orbits of their spacecraft to lessen the chance of collisions. Additionally, real-time monitoring data provides for a full knowledge of the behavior and distribution of space debris, assisting in the development of effective mitigation techniques.

Moreover, space debris laws contain recommendations for spacecraft disposal. Operators are expected to establish end-of-mission plans for their satellites and launch vehicles, ensuring that they are either safely deorbited or relocated to higher, less crowded orbits. This method helps reduce the accumulating of trash in widely utilized orbital areas, especially in low Earth orbit (LEO) where the density of objects is greater. Compliance with disposal criteria is vital in avoiding future accidents and limiting the long-term effect of space debris on the space environment.

Additionally, international collaboration is a vital part of space debris rules. The space environment is a global commons, and the actions of one country might have ramifications for others. Therefore, international agreements and treaties are vital for creating uniform standards and norms. Initiatives like the Inter-Agency Space Debris Coordination Committee (IADC) and the United states Committee on the Peaceful Uses of Outer Space (COPUOS) play essential roles in encouraging cooperation among spacefaring states and creating rules for space debris reduction and management.

Furthermore, culpability and responsibility are essential components of space debris legislation. In the case of a collision involving space debris, it is vital to establish responsibility and identify who is accountable for any subsequent damages. International conventions, like as the responsibility Convention, establish a framework for assigning responsibility and ensuring that operators take sufficient efforts to avoid accidents and limit any hazards connected with their space operations.

In conclusion, space debris laws serve a key role in maintaining the safety, sustainability, and long-term profitability of space operations. Through monitoring and tracking, disposal techniques, international collaboration, and responsibility frameworks, these policies handle the complex difficulties presented by space debris. As the amount of objects in orbit continues to expand, adherence to these principles becomes more critical in preserving our continuing exploration and exploitation of outer space.

15.3 International Cooperation and Governance in Space Activities

International collaboration and governance in space operations have become key components of the contemporary space age. As the exploration and exploitation of outer space continue to develop, the necessity for organized frameworks and agreements has expanded in tandem. This examination digs into the relevance, problems, and probable future orientations of international cooperation and governance in space.

One of the key reasons for promoting international collaboration in space operations arises from the fundamentally global aspect of space exploration. Unlike terrestrial operations, space activities transcend national borders and include a multiplicity of players from other nations. Joint activities are vital in attaining shared objectives like as scientific discovery, resource use, and maintaining the long-term viability of space operations. Moreover, sharing information, technology, and resources across countries promotes a more efficient and cost-effective approach to space exploration.

However, the route to efficient international collaboration in space is not without its hurdles. National interests, political rivalry, and security concerns may occasionally inhibit cooperation. Striking a balance between securing private knowledge and enabling common advancement is a hard challenge. Additionally, the legal and regulatory frameworks regulating space operations need continual refining to handle emergent concerns like space debris management, spectrum allocation, and planetary preservation.

The cornerstone of international collaboration in space operations resides in multilateral agreements and treaties. The Outer Space Treaty, approved by 110 nations, is a crucial agreement that defines the essential principles for the exploration and utilization of outer space. It stresses peaceful usage, the prohibition of territorial claims, and the need to aid astronauts in crisis. Other agreements, such as the Rescue Agreement and Liability Convention, further outline roles and liabilities in specific circumstances.

In recent years, there has been a rising acknowledgment of the need for new governance systems to handle modern concerns. Private sector engagement in space operations, illustrated by businesses like SpaceX and Blue Origin, creates a new dynamic that demands legislative adaption. Initiatives like the Artemis Accords, coordinated by NASA, attempt to build a framework for ethical and sustainable lunar exploration, stressing openness, interoperability, and the protection of historic sites.

Looking forward, the future of international collaboration and governance in space operations will likely be defined by many main issues. Continued technical improvements, particularly the introduction of new launch providers and satellite constellations, will need nimble and adaptive regulatory frameworks. Additionally, the growth of economic operations beyond Earth orbit, such as asteroid mining and space tourism, will necessitate better legal guidance.

In conclusion, international collaboration and governance in space operations are crucial to the sustained success of humanity's aspirations beyond Earth. While problems exist, the rewards of joint efforts in space exploration are considerable. Through changing treaties, accords, and novel governance systems, the international community can manage the intricacies of space operations while guaranteeing the responsible and sustainable use of this shared frontier.

Conclusion

In the concluding chapter of "Celestial Mechanics: Navigating the Cosmos," we bring together the threads of our research into the complex dance of celestial bodies. Throughout this book, we've gone deep into the basic ideas given down by visionaries like Newton, Kepler, and Euler. From the stunningly basic principles regulating planetary motion to the intricacies of orbital maneuvers and perturbations, we've unlocked the secrets that allow us to traverse the universe.

One of the main lessons from this expedition is the amazing accuracy and predictability of cosmic physics. Through the rigorous application of mathematical rigor and processing capacity, we can map the paths of spacecraft with surprising precision. This accuracy has not only permitted unparalleled scientific investigation but also underlies the infrastructure that sustains contemporary life on Earth, from GPS systems to communication satellites.

Yet, with this accuracy is a fascinating realm of dynamical intricacy. We've investigated the nuances of perturbations and resonances, discovering the subtle interplays that may lead to chaotic behavior in multi-body systems. Understanding and exploiting these phenomena will be vital as we push the frontiers of space exploration, pushing farther into the universe and embarking on expeditions to distant planets.

As we turn to the future, the frontiers of celestial mechanics beckon with fresh problems and possibilities. Interstellar travel, Lagrange clouds, and the incorporation of quantum effects into our models are merely a few of the exciting ideas that await study. These activities will surely call for the joint efforts of scientists, engineers, and visionaries from throughout the world, reflecting the genuinely international character of space exploration.

In finishing our voyage through celestial physics, we are reminded of the immense consequences of our knowledge of the universe. It not only strengthens our capacity to explore and exploit space but also gives a viewpoint that encourages a greater awareness for the connectivity of our globe and the larger cosmos. As we continue to explore the universe, may we do so with a feeling of wonder, curiosity, and a dedication to the proper management of this cosmic frontier.

Refrences

1. Danby, J. M. A. (1992). Fundamentals of Celestial Mechanics (2nd ed.). Willmann-Bell.
2. Bate, R. R., Mueller, D. D., & White, J. E. (1971). Fundamentals of Astrodynamics. Dover Publications.
3. Murray, C. D., & Dermott, S. F. (1999). Solar System Dynamics. Cambridge University Press.
4. Vallado, D. A. (2007). Fundamentals of Astrodynamics and Applications (3rd ed.). Springer.
5. Tisserand, F. (1903). Traité de mécanique céleste. Gauthier-Villars.
6. Vallado, D. A., McClain, W. D., & Petersen, C. D. (2004). Fundamentals of Astrodynamics and Applications (2nd ed.). Microcosm Press.
7. D'Amario, L. A., & Ocampo, A. C. (1987). The Restricted Three-Body Problem: Analytical and Numerical Aspects. Springer.
8. Murray, C. D., & Dermott, S. F. (1999). Solar System Dynamics. Cambridge University Press.
9. Roy, A. E. (2004). Orbital Motion (4th ed.). CRC Press.

10. Brouwer, D., & Clemence, G. M. (1961). Methods of Celestial Mechanics. Academic Press.

11. Iorio, L. (2010). Introduction to Space Dynamics. Springer.

12. Montenbruck, O., & Gill, E. (2012). Satellite Orbits: Models, Methods, and Applications. Springer.

13. Curtis, H. D. (2013). Orbital Mechanics for Engineering Students (3rd ed.). Butterworth-Heinemann.

14. D'Amario, L. A. (1986). Restricted Three-Body Dynamics: Analytical and Numerical Solutions. Reidel Publishing.

15. Belbruno, E. (2004). Capture Dynamics and Chaotic Motions in Celestial Mechanics: With Applications to the Construction of Low Energy Transfers. Princeton University Press.

16. Miele, A., & Mancuso, S. (1987). Optimal Trajectories for Space Navigation. Birkhäuser.

17. Abell, P. A., & Brashears, T. (2006). Solar System Dynamics: Discrete Dynamical Systems Approach. Academic Press.

18. Beletsky, V. V. (1983). Introduction to Celestial Mechanics. Springer.

19. Kozai, Y. (1962). Perturbations of asteroids and comets due to planets. Astronomical Journal, 67, 591.

20. Brouwer, D., & Clemence, G. M. (1961). Methods of Celestial Mechanics. Academic Press.

21. Gooding, R. H. (1993). Analytical celestial mechanics. Academic Press.

22. Edelbaum, T. N. (1961). Fundamentals of Astrodynamics. Dover Publications.

23. Moulton, F. R. (1914). An Introduction to Celestial Mechanics. Macmillan.

24. Curtis, H. D. (2009). Orbital Mechanics for Engineering Students (2nd ed.). Butterworth-Heinemann.

25. Beletsky, V. V. (1983). Introduction to Celestial Mechanics. Springer.

26. Vallado, D. A. (2013). Revisiting Spacetrack Report #3. Aerospace Corporation.

27. Binney, J., & Tremaine, S. (2008). Galactic Dynamics (2nd ed.). Princeton University Press.

28. Brown, E. W. (1978). Elements of Spacecraft Design. American Institute of Aeronautics and Astronautics.

29. Lamberts, R. (2019). Introduction to Space Flight Mechanics. Springer.

30. Prussing, J. E., & Conway, B. A. (1993). Orbital Mechanics (2nd ed.). Oxford University Press.

31. Zeilik, M., & Gregory, S. A. (1998). Introductory Astronomy and Astrophysics (4th ed.). Saunders College Publishing.

32. Wisdom, J., & Holman, M. (1991). Symplectic maps for the N-body problem. The Astronomical Journal, 102, 1528.

33. Bate, R. R., Mueller, D. D., & White, J. E. (1971). Fundamentals of Astrodynamics. Dover Publications.

34. Danby, J. M. A. (1992). Fundamentals of Celestial Mechanics (2nd ed.). Willmann-Bell.

35. Moulton, F. R. (1970). An Introduction to Celestial Mechanics (2nd ed.). Dover Publications.

36. Flandro, G. A. (1978). The Galileo Messenger. Jet Propulsion Laboratory.

37. Farquhar, R. W. (1990). The Use of Resonant Orbits in the Exploration of the Solar System. Springer.

38. Prussing, J. E., & Conway, B. A. (1993). Orbital Mechanics (2nd ed.). Oxford University Press.

39. Curtis, H. D. (2013). Orbital Mechanics for Engineering Students (3rd ed.). Butterworth-Heinemann.

40. Vallado, D. A., McClain, W. D., & Petersen, C. D. (2004). Fundamentals of Astrodynamics and Applications (2nd ed.). Microcosm Press.
41. Izzo, D. (2010). Advanced Mission Design for Solar Sail Missions. Springer.
42. Belbruno, E. (1995). Capture Dynamics and Chaotic Motions in Celestial Mechanics. Princeton University Press.
43. Montenbruck, O., & Gill, E. (2012). Satellite Orbits: Models, Methods, and Applications. Springer.